APL-STAT

A Do-It-Yourself Guide to Computational Statistics Using APL

James B. Ramsey
New York University

Gerald L. Musgrave
University of Michigan

LIFETIME LEARNING PUBLICATIONS
Belmont, California

A division of Wadsworth, Inc.

In preparing APL-STAT we were fortunate to have the help of many friends and colleagues. Rather than attempt to explain their individual contributions we simply list their names and express our thanks to each of them: Bert Alexander, Alea Curtis, Dorothy Dixson, David Edelman, John Hause, Robert Hessen, John Kassionas, Jan Kmenta, Alexander Kugushev, Charles Moore, Thomas Gale Moore, Jan Musgrave, Richard W. Parks, Virginia Perry, Alvin Rabushka, Grace Ramsey, Shannon Ramsey, Robert Rasche, Bernard Scheier, Bert Schoner, Andy Silver, Barbara Snarr, and Mike Sullivan.

Printed in the United States of America

1 2 3 4 5 6 7 8 9 10—85 84 83 82 81

Library of Congress Cataloging in Publication Data

Ramsey, James Bernard.
APL-STAT, a do-it-yourself guide to computational statistics using APL.

Includes index.
1. Statistics—Data processing. 2. Econometrics
3. Mathematical statistics—Data processing. 4. APL (Computer program language) I. Musgrave, Gerald L., joint author.
II. Title.
QA276.4.R35 519.5′028′5 80-15016
ISBN 0-534-97985-8

Contents

Preface

Please Read This Before Reading the Text!

This book explains how to perform both simple and complex statistical calculations using APL. ''APL'' is an acronym for ''A Programming Language''—a computer programming language that is ideal for the computational work done in statistics.

The authors are both economists, and the content reflects their professional interests. However, political scientists, physicists, sociologists, industrial psychologists, public health and dental researchers, and others have used this book and found it helpful.

No previous knowledge of computers, computer programming, or methods involved in statistical computation will be needed to understand this book. You will start from the most elementary statistics and progress to more complicated procedures on a gradual step-by-step basis. The numerous examples, exercises, and statistical applications are drawn from a variety of fields. Emphasis is placed on how to obtain the statistical results with ease. Using this book you will be able to perform computations that otherwise would be so cumbersome or time-consuming that you would not do them. You also will be able to perform experiments and computer simulations with relatively little effort.

The APL statistical procedures presented are useful to researchers, analysts, managers, and anyone concerned with statistical calculations. We believe that when you have seen how easy it is to perform these computations, you will be as pleasantly surprised as we were. If you are familiar with computers here is a dramatic example of the simplicity of APL compared to the FØRTRAN statements used to compute the arithmetic mean. If you are a novice in these things don't be frightened—everything will be explained.

An Example of FØRTRAN and APL

FØRTRAN

```
      DIMENSIØN X (1000)
      READ (5,99)N
 99   FØRMAT (I4)
      READ (5,100) (X(I), I=1,N)
100   FØRMAT (9F8.0)
      SUM = 0.0
      DØ 10 J = 1,N
 10   SUM = SUM + X(J)
      AVE = SUM/N
      WRITE (6,20)AVE
 20   FØRMAT (F10.4)
      END
```

APL

```
X←⎕
⎕←AVE←(+/X)÷⍴X
```

To estimate the parameters of $Y = B_1 + B_2 X_2 + B_3 X_3 + \cdots + B_N X_N + u$ via multiple regression, you could type in APL:

Use of ⌹ in Multiple Regression

```
B ← Y⌹X
```

In other computer languages an equivalent program might take 50 statements.

This book is not just an introduction to APL programming, although many people have learned APL from it. Certainly it is not a statistics textbook, but readers have commented that they never really understood certain statistical concepts until they "tried real numbers to see how the formulas worked." This book is a valuable aid to understanding statistics because it actually computes results and even displays probability distributions graphically. By the time you finish you will know a lot about APL programming. And after you spend a few hours at the computer, you will find that it is easier to program your own work than it is to learn to use the "canned" (FØRTRAN) routines available at the computer facility. More importantly, you will understand what you are doing and how the results are obtained. We have long maintained that the less you are asked to accept unquestioningly, the better is your intellectual health and the greater will be your interest in statistical subjects.

This book is not primarily a textbook. It is a book for the person who understands basic statistics, who wants a painless way to compute results, and yet wants to know what is really going on. We think that teachers of basic or applied statistics and especially econometrics will find our approach using APL to be an important part of a practical statistics course. Students are often assigned "artificial," "theoretical," or "academic" problems, situations, and exercises. These assignments are not made because the instructor thinks such things are important. Actually, most instructors understand the difficulty of tackling *real* statistics problems. Consequently, when the amount of computational pain the student (and teacher) must go through to get the statistical result is compared to the "statistics" that can be taught, a stress on pure theory almost always results. Thus, after a course (or even several courses), an individual may be unprepared to solve the first problem—how to perform the calculations! The use of APL minimizes these difficulties.

We think that when you complete APL-STAT you will agree—programming can be easy!

Because the text proceeds in a carefully structured sequence it is important that you follow it exactly and that you make sure you thoroughly understand each section before moving to the next. Later sections assume that prior sections have been mastered. You should do the exercises and check your answers in the back of the book. Above all, you can teach yourself a lot by experimenting, so try it.

Purpose of These Comments

If you forget something, the primitive function glossary at the back of the book will help you recall earlier material. If you need more information, the side of most pages has brief comments. These comments contain the name and symbol of the APL operator introduced on that page. You will be able to flip through the book quickly and locate what you want, using the comments. They also provide a quick visual guide to the major topics in any section.

We have a request. In the back of the book is an error sheet for recording *our* omissions, bad language (though never foul!), and other sins. We would be most obliged if you would send us this error sheet with your comments. The next edition will then be much better with your help.

JAMES B. RAMSEY GERALD L. MUSGRAVE

Note to Instructors

Instructors can assign much more meaningful examples and exercises using the procedures in this book than using either canned programs or hand calculation. Students will not be spending time in tedious calculation or in using the computer as a black box. Students will be able to perform calculations, including complex matrix algebra, know how they are done, and see the numerical results. They will be able to obtain results they understand. One example is where a multiple regression model requires the intercept to be ''forced'' through zero. It is surprising how simple the mathematics of this is (not having a column of ones in the regressor X matrix). It is also surprising how few preprogrammed packages allow this option. In APL you can modify your program to handle this change in a matter of moments.

Computer simulation and generation of distributions become a relatively trivial task in the hands of an APL-proficient student. We could enumerate a long list of such examples, and once you start you will see them too. Also, we have included our benchmark program data on the Longley regression problem in Appendix B. You may find it interesting to compare the computational accuracy of APL programs with the canned ones on your home computer or at your computer center.

In using this book as a text you might consider the following ideas. The titles of certain sections, e.g., *The Normal Distribution* in Chapter 6, are starred. These starred sections involve mathematical material which may be beyond the scope of an elementary course in statistics that doesn't have a mathematical prerequisite. Any APL instructions introduced in such sections will not be used anywhere else in the text without reexplanation. So starred sections can be dropped without fear of losing some important information about APL.

The book is carefully structured in that it follows the usual pattern of topics in the introductory statistics course and only uses as much APL as is needed to get the job done. Consequently, it is important that, except for the starred sections, the sequence be followed and sections are not skipped.

If you decide to alter the presentation of statistical subjects, have your students read the APL-material in sequence, even if they skip the earlier presentations of the statistics. A number of readers have used this approach and found it to be satisfactory. In these cases the readers either knew statistics or were not interested in statistics per se. They wanted to learn APL and found this approach to be effective. One reason for this is that APL instructions are introduced to solve specific problems rather than presented in the abstract.

Each chapter has a large number of exercises and applications. The exercises help in exploring the use of APL concepts, functions, and symbols. The statistical applications help extend the depth and breadth of APL use. Throughout the book, experimentation is encouraged to expand and intensify interest and understanding.

An elementary nonmathematical course in statistics would usually stop at Chapter 9, which covers contingency tables, analysis of variance, and simple linear regression with one regressor. Chapters 10, 11, and 12 introduce various aspects of matrices and prepare the way for multiple linear regression analysis and topics that might be regarded as more "econometric." You may find that the use of APL will allow you to cover Chapters 10 through 13 as well. This is important since the rudiments of matrix algebra can be taught quickly using APL. The benefit will be that you can enable your students to master multiple linear regression and more complicated analysis of variance techniques more easily.

Three administrative matters might be of interest. Many computer centers have only a few APL terminals. Don't let this apparent difficulty slow you down. First, if the terminals use a typing ball or a daisy wheel, the center can obtain APL balls or print wheels. They are easy to switch, are low in cost, and small adhesive labels are available for the keys. Second, if the terminals use a non-APL matrix printer or if the terminals are CRT's without APL characters, another solution is available. A Mnemonic character set that substitutes for the APL symbols is available. The multiple regression example in the preface was coded as

```
Y←Y⌹X
```

using the standard APL character set. In the Mnemonic character set it would be written as

```
Y←Y.DQX
```

Appendix C contains both the standard and Mnemonic character sets. Third, some computers have implemented only the monadic version of domino. In this case you simply enter the following two lines

```
∇YDQX
(⌹((⍉X)+.×X))+.×((⍉X)+.×Y)∇
```

when you enter *YDQX* the result is the same as if *Y⌹X* had been entered.

If in using the book you have any comments that would be helpful to others please pass them along to us and we will incorporate them in the next edition.

1

Introduction

1.1 Overview of APL

APL is a powerful and versatile computer programming language. When you use this language to communicate with the computer it will be as if you were personally operating the machine. APL is designed to operate on small microcomputers no larger than a typewriter, on minicomputers the size of one or two office desks, and on large maxicomputers the size of a truck. No matter how large or small the computer, once you log-on to the system it will appear from your perspective that you have a one-to-one relation with the computer. The APL contained in this book has been used on micro-, mini-, and maxicomputers produced by a variety of manufacturers. We found the APL language to be remarkably similar in all of these cases.

Administrative Procedures

The procedures used to log-on to the various systems that we have used vary greatly. Each computer center has its own administrative procedures, keywords, passwords, and account verification methods. In addition, you usually need to connect your computer terminal to the computer itself and this process can be mysterious at first. There is really nothing to this at all. Nevertheless, sometimes people who hang around computer centers make a big deal about the administrative and technical matters surrounding the use of the machine. The truth of the matter is that the procedure is much the same as getting a key for an office, registering for a class, or signing up for Little League. It's a hassle. Every organization thinks that there is only one way to do it, and yet every way is different. Appendix A contains a brief description of how it is done at the Stanford and NYU computer centers, and on an IBM 5120 desk-top computer. This description should allow you to understand better the procedures that are used with your computer. In a short time the mystifying intricacies of gaining access to the computer become second nature. You type a few words and numbers and you are ready to go.

The APL Keyboard

We have included a few diagrams of typical APL keyboards in Appendix A. The alphabetic characters are in exactly the same position as they are on a standard typewriter. These letters are all capitals but (wouldn't you know it) they are in the lowercase positions. Holding the shift key down while pressing a specific key enters a special APL symbol. Each of these symbols performs a specific operation in APL. As you can see, the keyboards are almost identical, and in the very few instances where some minor differences do exist we will explain them. One of the most frightening things that the new APL programmer encounters is the APL character set. All those strange symbols are indeed foreboding. However, our experience has been that the symbols are easy to learn. They are not much more difficult to learn than the international road signs, especially if you take them one at a time in the context of an actual problem.

Some General Features of APL

CLEAR WS

Now suppose that you are sitting in front of the keyboard and you have logged-on. The computer has responded with the message *CLEAR WS*. The computer is indicating that you have been allocated a part of the computer—APL calls it a Work Space—named *CLEAR*. Now you can communicate with the computer, and it is in fact much like an electronic hand-held calculator except that it is much more powerful. To turn off the computer you simply type *)OFF*, for example, and log-off. You will soon see how APL can be used as a very powerful calculator in the immediate execution mode. However, it can do much more.

)OFF

You can define a set of instructions that will perform tasks such as balancing a checkbook; computing means, standard deviations, and regression coefficients; or directing the computer to simulate a Las Vegas casino game. In APL, the set of instructions is called a defined function. After the function has been defined you simply refer to it by name. The same instructions, operating on different data, can be used over and over again.

State Diagram

Figure 1.1 is a state diagram that represents these three APL modes. When you log-on you are given a clear work space, you are in immediate execution mode, and you have a powerful calculator at your disposal. You can enter data, process the data with a one-line APL expression, define an entire new work space with different functions and data, and test your functions on a line-by-line basis before you program the whole set of instructions.

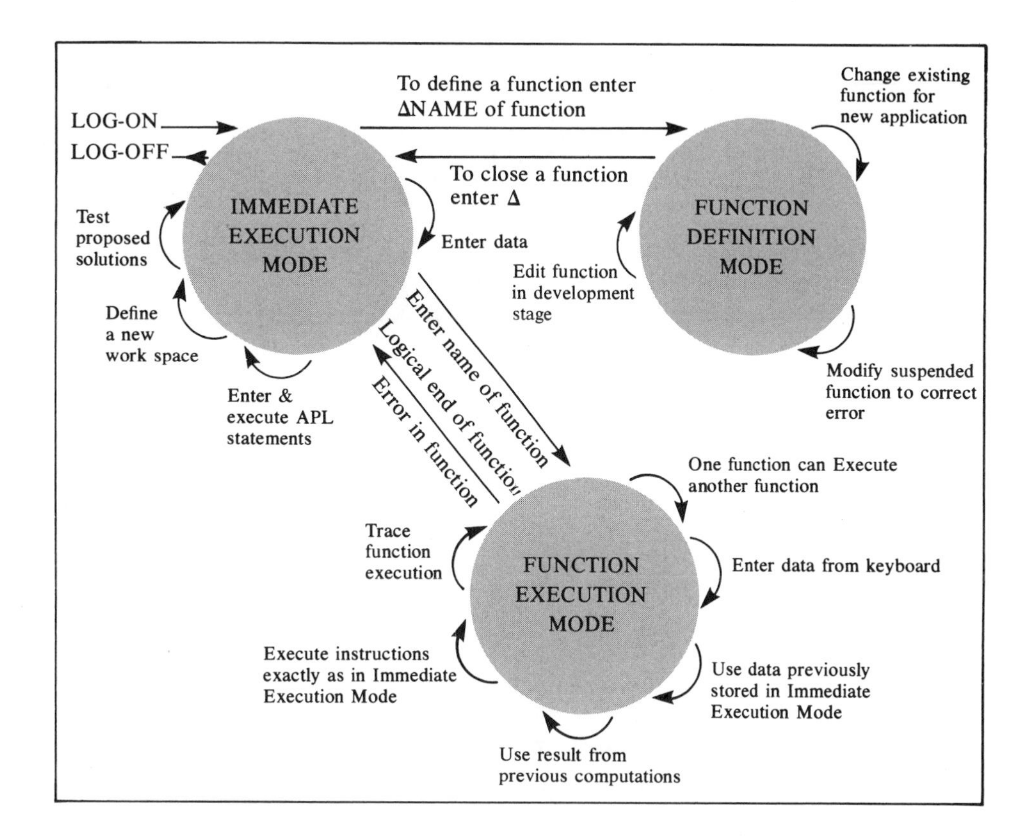

Figure 1.1

Then you can define your own function, edit any part of it, or modify it for a particular application. Also, should a function stop because of a programming error and further processing thereby be suspended, you can correct the error by editing the function and then resume the function's execution from the point of suspension. You need not start from the beginning if your previous calculations were correct.

A function is executed by simply entering its name. You can specify the particular data set to be processed, and your function can call other functions, request data, and produce results for use by other functions. In addition, you can trace the execution of your function by having the results of any line or group of lines displayed—all of this without having to write any output statements. When your function's execution is completed it returns you to immediate execution mode where you began. We hope that this sounds simple, straightforward, and like something you can do—because it is!

1.2 Road Map of Where We Are Going and How We Will Get There

In the next chapter you will learn how to use APL as a calculator. After these basics are under your belt, the general presentation is to explain a statistical problem and then to solve it using APL. On the way to the solution the various APL functions and programming methods are presented and explained. We first discuss the sample mean and median, stan-

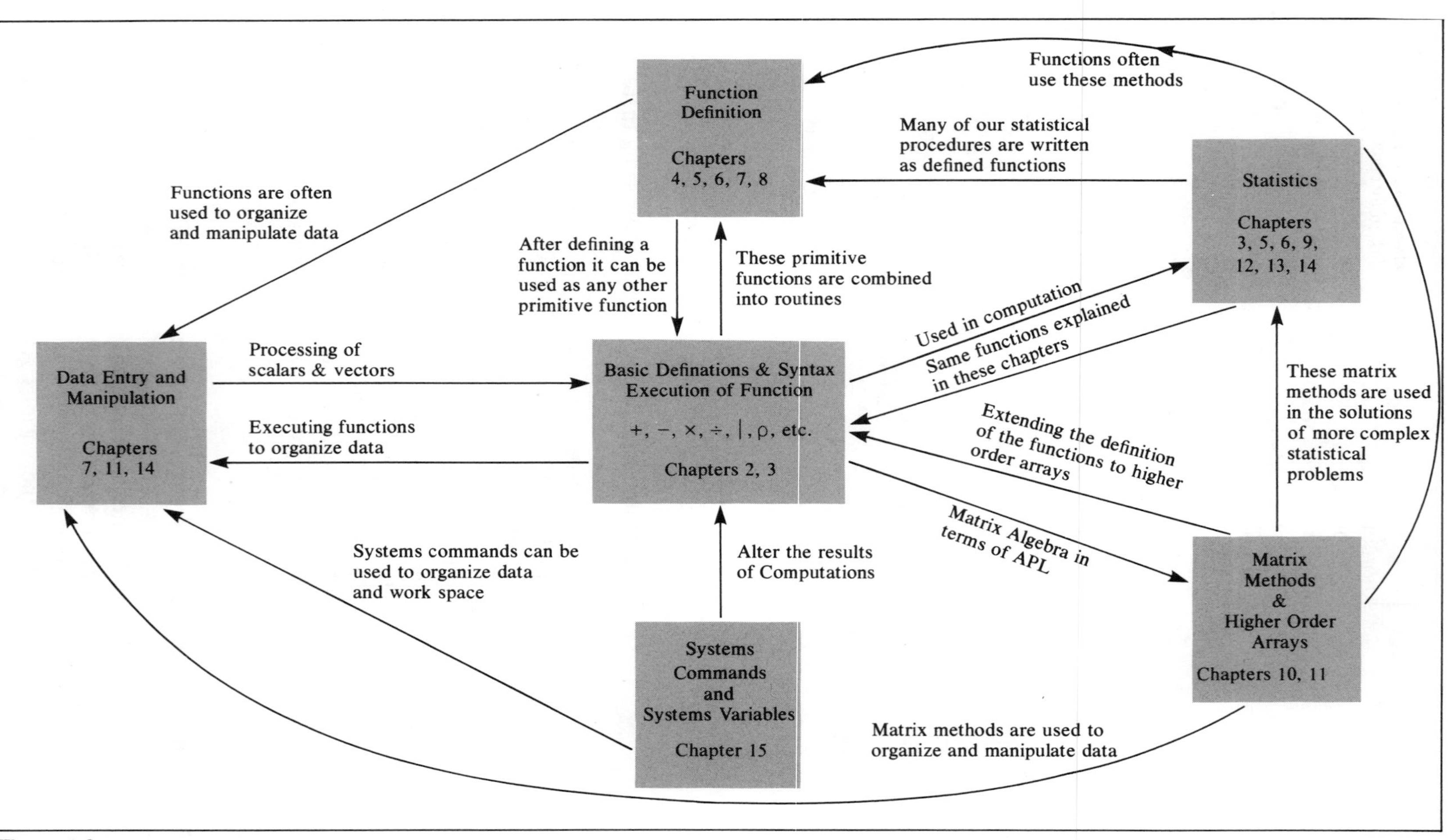
Function
Definition
Chapters
4, 5, 6, 7, 8
Functions often
use these methods
Many of our statistical
procedures are written
as defined functions
Statistics
Chapters
3, 5, 6, 9,
12, 13, 14
Functions are often
used to organize
and manipulate data
After defining a
function it can be
used as any other
primitive function
These primitive
functions are combined
into routines
Used in computation
Same functions explained
in these chapters
Data Entry and
Manipulation
Chapters
7, 11, 14
Processing of
scalars & vectors
Basic Definations & Syntax
Execution of Function
+, −, ×, ÷, |, ρ, etc.
Chapters 2, 3
These matrix
methods are used
in the solutions
of more complex
statistical
problems
Executing functions
to organize data
Extending the definition
of the functions to higher
order arrays
Matrix Algebra in
terms of APL
Systems commands can be
used to organize data
and work space
Alter the results
of Computations
Matrix
Methods
&
Higher Order
Arrays
Chapters 10, 11
Systems
Commands
and
Systems Variables
Chapter 15
Matrix methods are used to
organize and manipulate data

Figure 1.2

dard deviation, covariance, and higher order moments. Then we investigate a number of the most prominent statistical distributions, including the binomial, Poisson, and normal density and cumulative distribution functions. After you learn how to handle more complex data structures in APL and to write more general and powerful functions you will learn how to diagnose and correct programming errors. After you go through a case study using APL in a research project, we present an introduction to elementary linear correlation and regression, analysis of variance, and the chi-square and F distributions. Next we show how to do matrix algebra in APL, including the operation of matrix inversion, which is performed with one symbol, ⌹. Multidimensional arrays are discussed in Chapter 12, where the various APL functions are explained in relation to these higher order arrays. The final chapters concentrate on computational statistics related to multiple linear regression, two-stage least squares, instrumental variables, Aitken estimators, Durbin's First Order Autoregressive Models, and K-class estimators including limited information maximum likelihood estimators.

Don't let this impressive sounding jargon put you off. The first half of the book has been understood by good high-school students, and they were able to write APL programs after only a few hours of study. The later chapters have been used in both undergraduate and graduate classes. Also, the statistical routines have been used by a number of our colleagues in their statistical research. So you can see that while much of the material is technical, it progresses at a measured rate. Figure 1.2 is a schematic representation of APL-STAT. It might help you to visualize how the various components of APL are related.

We can summarize our position this way:

APL
TRY IT—YOU'LL LIKE IT

So turn the page and let's go . . .

2

Getting Started

2.1 Some Keying Conventions

Now that you are seated comfortably in front of your terminal or minicomputer, everything is switched on, and the terminal is set to receive your instructions in APL, we can begin. Our first task is for you to gain some familiarity with the use of your keyboard as shown in figure 2.1.

Sometimes we want to indicate to you very clearly that there is a blank space. For example, consider the character string ABCD EFG, which you type by hitting the A, B, C, D keys, the space bar, and then the keys E, F, and G. Blank spaces will be indicated, but only when we need to stress that there is a blank, by printing an ampersand (&) in a subscript position. In the above example, we would print $\text{ABCD}_{\&}\text{EFG}$. You can read this as: A, B, C, D, and E, F, G. You will not find the & character on your APL keyboard; we use it in the earlier chapters to emphasize blanks until your eye is accustomed to the idea.

2.2 Simple Arithmetic

Arithmetic Functions +,-,×,÷

We will start by making sure we know how to add (+), subtract (−), multiply (×), and divide (÷) numbers on the computer. The symbols +, -, ×, and ÷ are the symbols for the mathematical operations of adding, subtracting, multiplying, and dividing, respectively. On the IBM 5120, for example, they are found on the far right-hand side of the keyboard, next to the keys with the integers from 0 to 9. You may also use the numbers shown on the top row of the main keyboard and the arithmetic functions shown at the end of that row (this is the most common configuration).

You instruct the computer to perform a calculation by hitting the RETURN or EXECUTE key; the instruction is executed only *after* you hit the key.

Figure 2.1 IBM 5120 desktop computer showing the APL character set, numeric pad, and special function keys.

This photograph showing a 5120 desktop computer can be programmed in either Basic or APL with the flip of a switch. The keyboard is exactly like a standard typewriter in that pressing the shift key (either of the keys with the wide arrows on the bottom rank of keys) results in the APL characters being entered into the computer. A convenient feature is that by holding the command key (CMD, on the far left) and pressing one of the keys on the top row will produce an entire command. For example, holding down CMD and pressing 1 results in the command)LOAD being entered automatically.

IBM 5120 showing keyboard characters that can be entered using the command key.

This photograph shows the special overstruck characters that can be produced with one stroke. The command key is held down and any of the individual keys now represents a new symbol or combination of key strokes. For example, pressing the CMD key and the F key results in the divide quad or domino function being entered. If the machine were in the Basic programming mode the characters INPUT would have been entered. Using the CMD key saves a number of key strokes and is a handy feature.

Addition

Add +

```
      1+2
3
```

If you did not get the answer after keying in the digit 2, hit the RETURN or EXECUTE key. Now try addition with decimals:

```
      1.2+0.6
1.8
```

But if you key in

$1._{\&}2+0._{\&}6$

```
1 8
```

This does not look right! What happened?

Clearly, embedded blanks in real numbers (numbers that include a decimal) cause problems! So do not embed blanks in real numbers. You will understand why you got $1_{\&}8$ and not `1.8` by the end of this chapter.

Try

```
      1+
```

SYNTAX ERROR

```
SYNTAX ERROR
      1+
      ∧
```

You have made an error and the symbol ∧ (a caret) marks the point at which the error occurred. Unfortunately, we all get much too familiar with this symbol! The error was called a syntax error because the statement "execute `1+`" is ungrammatical; it does not make sense to tell the computer to "add 1." The computer's response is to say, "Add 1 to what?" or "How can I do this?"

Subtraction

Minus -

Key in `3`, minus sign (`-`, which is next to the plus (+) sign), and `2`, then EXECUTE:

```
      3-2
1
```

or

```
      2-3
¯1
```

Notice that on the last response negative 1 was printed by the computer as `¯1`. The superscript negative indicates the negative sign of the number and must be carefully distinguished from the − in −1. In the latter case the symbol `-` represents the operation of subtraction. How do we know the difference? By position. For example,

Negative Numbers

`¯2` represents the *number* "negative 2";

`-2` represents the operation of subtracting 2 from whatever is to the left of `-`.

How do you type the number "negative 2"? This is done by typing the symbol `¯`, which is upper-shift `2` on the keyboard. Try it.

```
      ¯2
¯2
```

(Remember to hit EXECUTE!)

Now use the minus operator symbol key:

```
      -2
¯2
```

What happened here? `3+` gave a syntax error, why didn't `-2`? The answer is that APL interprets the operation `-2`, when nothing is on the left, as the instruction "Make the argument (i.e., whatever is on the right) the negative of whatever it is." Try `-¯2` and `--2`. You get `2` in both cases. Try the following as well:

```
      +2

      +¯2

      -+2

      2-
```

So if the `+` or `-` functions are on the left of a number, the sign of the number is unchanged by `+` and reversed by `-`. But if the *number* is on the left of the *function,* you will get a syntax error.

Multiplication

Multiply ×

```
      2×3
6
      3.5×2
7
```

Division

Try

Divide ÷

```
      6÷3
2
      5÷2
2.5
      0÷4
0
      4÷0
```

DOMAIN ERROR

```
DOMAIN ERROR
      4÷0
       ^
```

We have hit another error. A useful mathematical convention is that division by zero is an undefined operation, and that is what the computer is telling you. In this case the syntax or "grammar" of the use of the function ÷ is correct, but the operation cannot be performed with the number `0`; `0` lies outside the domain of validity for the operation of division. But what about `0÷0`? Try it.

```
      0÷0
1
```

Without going into details at this stage, merely note that this is another useful convention—in short, an agreement as to what to do with such an operation.

2.3 Arrays

Arrays

We will now introduce you to the single most important aspect of the APL language, the array. An array is an ordered arrangement of numbers or characters. A simple example is a linear arrangement of numbers, such as `1 3 2 4 5`, or characters, such as *AN&ARRAY*. In one form or another, arrays play a vital role throughout this book. Try the following:

```
      3+263
266
```

and

```
      3+2&6&3
5 9 6
```

Blanks

What has happened here? In the second example `2 6 3` is treated as a list or array of three numbers, viz., `2`, `6` , `3` in that order. So an array of numbers is created by separating each number in the array by a blank. Another way to do it is to key in `2,6,3` where the comma, instead of blanks, separates the individual digits. If you recall the comment made above about blanks inside real numbers, you will see that it is dangerous not only for you to embed blanks in real numbers in an array, but also for you to embed blanks within integers as well.

Try the following:

```
      3+1.2&2.3&3.3&4.1
4.2 5.3 6.3 7.1
      3+1.&2&2.3&3&.3
4 5 5.3 6 3.3
```

If these results seem strange (or if you do not get either result) carefully check your keying of numbers *and* blanks. In the second example, the result shown occurs because `3` is added to `1.`, `2`, `2.3`, `3`, and `0.3` in turn;

the blanks denote from the right where one number ends and the next begins. Also try

```
        3+1.2,2.3,3.3,4.1
4.2 5.3 6.3 7.1
```

APL and Arrays

or

```
        3+1.,2,2.3,3,.3
4 5 5.3 6 3.3
        1&2&3+3
4 5 6
```

Notice here that just as you can add a number to a list of numbers you can also add a list of numbers to a single number.

```
        3+1&&2&&&364
4 5 367
```

So much for the blanks. Now let us get back to the main issue: What is meant by adding 3 to an array of numbers? Quite simply, and as you would expect, 3 is added in turn to each of the numbers in the array. Now try

```
        4×2&3&4
8 12 16
        4÷2&4&8
2 1 0.5
        2&4&8÷4
0.5 1 2
        3&2&1-2
1 0 ¯1
```

APL Functions and Arrays

The general rule we see is that for any function f such as +, -, ×, or ÷, a number n, and an array $a_1, a_2, \ldots, a_p$, the statement "nf array" produces an array $nfa_1, nfa_2, \ldots, nfa_p$, and the statement "array fn" produces an array $a_1fn, a_2fn, \ldots, a_pfn$.

If one array lets you do a series of operations all at once, what will happen if you use two arrays? Try

```
        1&2&3+1&2&3
2 4 6
```

Clearly, each element of the first array is added to the *corresponding element* of the second array. Similar results hold for the other arithmetic operations.

But what if the two arrays have different numbers of elements in them (i.e., what if the arrays have different lengths)? There will be some elements in one array to which there are no corresponding elements in the other. So if we try to operate on arrays of different lengths, we get a

LENGTH ERROR

LENGTH ERROR. Try

```
    1&2+1&2&3
LENGTH ERROR
    1 2 + 1 2 3
          ^
```

But the following is fine:

```
    1&2&0+1&2&3
2 4 3
```

Summary

+ , -, ×, and ÷ are *arithmetic functions*.
"3+2" adds the numbers 3 and 2; "3-2" subtracts 2 from 3; "3÷2" divides 3 by 2 : "3 × 2" multiplies 3 times 2.
"+ number" returns the number.
"- number" changes the sign of the number.

An *array* of numbers is formed by entering numbers in a list separated by blanks or by commas. We represent blanks where necessary in this text by &. More complex arrays will be discussed later.

Numbers, arrays, and the arithmetic operations that we discussed in this chapter can be combined as follows:

Number *f* Number	yields a number.
f Number	yields a number.
Number *f*	yields Syntax Error.
Number *f* Array	yields an array.
Array *f* Number	yields an array.
Array *f* Array	yields an array only if the arrays have the same number of elements (the "same length"). It yields Length Error when the arrays have different lengths.
f Array	yields an array.
Array *f*	yields Syntax Error.

Exercises

APL Practice

Let's explore the use of the functions defined in this chapter:

1. (a) +2 positive two

(b) 2+ Syntax Error

(c) `1÷2` one divided by two
(d) `-2` minus two
(e) `¯2` negative two
(f) `-¯2` minus negative two
(g) `¯-2` the negative of minus two
(h) `÷0` Domain Error
(i) `3÷0` Domain Error
(j) `3÷(2-2)` Domain Error
(k) `3+¯2` three plus negative two
(l) `3-+2` subtract positive two from three
(m) `3×¯2` multiply three by negative two
(n) `3×÷¯2` three times the reciprocal of negative two
(o) `3×¯÷2` three times the negative reciprocal of two
(p) `-3×÷¯2` the negative of the answer to (n)
(q) `-3÷¯2` the negative of three divided by negative two

2. You can get a better idea about the use of arrays by trying the following exercises:

(a) `1&2&3+2` add a number to an array
(b) `1&¯2&3-2` subtract a number from an array
(c) `1&¯2&3+¯1&2&¯3` add two arrays
(d) `(1&¯2&3)-(¯1&2&¯3)` subtract two arrays
(e) `1&¯2&3-¯1&2&¯3` subtract two arrays
(f) `1&¯2&3×¯1&2&¯3` multiply two arrays element by element
(g) `1,¯2,3×¯1,2,3` same as (f)
(h) `(1&¯2&3)÷(¯1&4&¯3)` divide one array by another, element by element
(i) `1 ¯2 3÷¯1 4 ¯3` same as (h)
(j) `1&2&3×-1&2` Length Error
(k) `1&0&3×÷1&0&3` Do you get Domain Error? Why or why not?
(l) `1-&¯2&3+2` add the number two to the array `¯2 3` and subtract the sums from one

Statistical Applications

1. What is the arithmetic average of 10 and 20?
2. What is the reciprocal of the arithmetic mean of the reciprocals of 10 and 20? Verify that this number is smaller than the arithmetic average of 10 and 20.
3. Find the volume of a cube whose sides are 4.5 ft long.

4. The following three measurements were taken on one side of a cube: 0.00000060, 0.00000065, 0.00000063. What is your estimate of the volume of the cube?

5. Seven prices for a popular 35 mm SLR camera were collected from a recent photography magazine: $259.95, $245.00, $254.99, $259.99, $259.95, and $249.95.

   a. Compute an average price for the camera.

   b. What is the range of prices?

   c. How much more is the highest price than the average price?

   d. What would be the percent saved by purchasing at the lowest price compared to the highest price?

6. A local business selected a representative week's returned checks due to insufficient funds or fraudulent accounts. The checks were written for $23.41, $184.24, $73.12, $2.48, $32.00, $14.28, $58.61, $84.00, $41.41, $83.27, and $102.87. What would you forecast the yearly total amount of returned checks to be?

3

Some Elementary Statistics

If you have read the first few chapters in any book on statistics or econometrics, you will have noted that the sample mean appears quite prominently. In fact, if you continue using statistics you will be computing a large number of means. It will save a lot of time if we can discover a quick way to get the computer to do it. Before tackling our first statistic, we have to learn an important fact about how a computer reads instructions in APL.

3.1 The Computer Reads from the Right

In order to compute a mean, we need an array of numbers and a knowledge of how many elements (numbers) it contains. Suppose we have the array (1 2 3 4), which obviously has four elements in it, and we want to calculate its mean. Mathematically, the operation can be written as:

$$(1 + 2 + 3 + 4)/4 = 2.5$$

In APL we can enter the following statement:

```
      (1+2+3+4)÷4
2.5
```

Great so far. But suppose we entered:

```
      1+2+3+4÷4
7
```

Computer Reads From Right to Left

We made another mistake! But this one is a very, very important one to remember. In APL a string of mathematical operations is carried out *from right to left*. Since we are accustomed to reading from left to right, you can see that until you are used to the idea, you can make some bad mistakes. Indeed, for the next few chapters you are strongly advised to practice reading all the computer statements from right to left.

Consider the first example: (1 + 2 + 3 + 4) ÷ 4. The computer does the following. Starting from the right, the computer recognizes a number, then a function requiring two arguments, such as ÷, ×, -, or +, then a right parenthesis. This parenthesis tells the computer to keep going to the left until it encounters a matching left parenthesis; then whatever is contained between the left and right parentheses is to be divided by four. Within the parentheses the computer recognizes a 4, the function, +, and then the number 3. It performs the operation 3 + 4 and stores the result. Proceeding to the left, it recognizes another function symbol, +, followed by another number, 2, so it adds 2 to (3 + 4), and so on.

All of this is simple enough, so let us try a trickier example. Do this one by hand first and then check your result on the computer.

```
      1+2-3-4+5-6-7
10
```

If you got ¯12 instead of 10, then that is exactly what we wish to explain. If we add parentheses, the above expression can be written as

```
      1+(2-(3-(4+(5-(6-7)))))
10
```

In case you haven't got it yet, the following table should help:

Operation Number	Operation	Result
1	6-7	¯1
2	5-¯1	6
3	4+6	10
4	3-10	¯7
5	2-¯7	9
6	1+9	10

3.2 Two Arguments or One?

A few paragraphs back we said that the functions ÷, ×, -, and + require two arguments, but in Chapter 2 we successfully used the + and − functions with only one argument, provided that the argument was on the right of the function, not the left. At the moment all of this may be confusing, but it won't be after we show you how useful it is to have a function that can take either one or two arguments.

Monadic Functions
Dyadic Functions

First, a little terminology in case you dip into an APL manual or talk to a programmer friend: functions that take two arguments are said to be *dyadic,* and those that have one argument are *monadic;* `1+2` is a dyadic use of +, `+2` is monadic. The *symbols* for most functions are used to represent both an operator that is dyadic and one that is monadic—two functions for the price of one symbol!

For example, the function ÷ can be used in two ways:

Reciprocal ÷

Monadic function: Symbol: ÷, function: reciprocal
Example: `÷2`

```
.5
```

Dyadic function: Symbol: ÷, function: division
Example: `4÷2`

```
2
```

In the first case the symbol ÷ indicates the reciprocal (or 1 ÷ argument); in the second case the symbol ÷ indicates the operation of division ($argument_2$ divided by $argument_1$).

The symbol − represents two functions—the monadic function of arithmetic negation (more simply, "changes the sign") and the dyadic function of subtraction. The symbol + represents addition in its dyadic form; in its monadic form it preserves the identity of the argument, i.e., + number returns the number itself. The symbol × is used for both the dyadic function of multiplication and for the monadic function "signum," which will be mentioned later.

Recall that with monadic uses the function comes *first,* then the argument. A number followed by a function and nothing else gives a *SYNTAX ERROR*.

SYNTAX ERROR

In using symbols that can represent different functions depending on whether they are being used monadically or dyadically, remember to read from the right. After a little more practice in the exercises you will soon find no difficulty in distinguishing monadic from dyadic uses of functions.

3.3 Variables and Assignment

If we want the mean of the array 1, 2, ¯3, ¯4, 5, ¯6, ¯7, what should we do? Typing out (1 + 2 − 3 − 4 + 5 − 6 − 7) ÷ 7 is incorrect; try it and you will see. (Remember to read from the right, performing each function in turn and storing the result.) Well, there is a very easy solution, but first we will find it useful to give arrays and scalars (a scalar is a single number) names, so that when using the array we can refer to the name instead of writing out the whole array each time. This procedure is called "assignment." Assignment uses the key next to the *P* key. (Do not confuse this key with the shift control keys, which also have arrows on them. The latter keys are used for editing by moving text to the right or left, up or down. On the IBM 5120 they are located next to the ATTN key; on other keyboards they are usually on the right next to the number keys or the numeric key pad.) Type out

Assignment

```
X←1&2&¯3&¯4&5&¯6&¯7
```

and hit the EXECUTE key. Nothing seems to happen. Try typing *X* and hit the EXECUTE key:

```
      X
1 2 ¯3 ¯4 5 ¯6 ¯7
```

Success! We now have the array we want stored in the computer with a name, *X*. "Executing" *X* tells the computer to print out or display *X*. Try

```
      N←7
      N
7
```

Variable Names Valid/Invalid

Letters of the alphabet together with numbers, but only after the first character, can be used to define names of arrays or scalars. Special symbols for operations, spaces, punctuation marks, and so on cannot be used. Some examples of valid and invalid names are:

Valid Variable Names	Invalid Variable Names
A	3A
ABLE	÷
B3C1	¯A
Z	A'
Z̲	(B)
A_OR_B	+C
	A*

Z̲ is created by typing *Z*, backspacing once, and hitting the upper shift *F* key. *Z* and Z̲ are different names. *A_OR_B* is keyed by typing upper shift *F* for _. Keying in an invalid variable name with assignment produces a syntax error.

Results can be lost when you log-off

An important question arises at this point. If someone defines a number of variable names by assigning values to them, what happens when he signs off the computer or turns off the power on his minicomputer? As one might suspect, all is lost! However, we will learn in Chapter 7 how to save important material for use at a later time. For now, *remember that if you log-off after having assigned values to variables, the variables will not be defined when you log-on next time.*

3.4 A System Command:)VARS

System Command

An aid in this regard is the system command)*VARS*. First, we have to define a system command. This is an instruction to the computer concerning the manner in which it carries out your APL instructions; system commands are rather like sending instructions to an operator who is keeping a constant record of all that you do on the computer. System commands are easily recognized; they all start with a), a right parenthesis.

)VARS

The use of)VARS will illustrate the idea. Suppose, after a long session on the computer, you have forgotten which variables you have defined. An obvious idea is to ask the computer what variables you have used. But it is clear that we need some way to make sure the computer knows we are asking a question about the system and how it is operating, and that we are not making another statement in our calculations. In APL, the distinction is very simple: system commands begin with a right parenthesis,), which is keyed as upper case]. For example, typing

```
      )VARS
N        X
```

instructs the computer to give us a list of the variable names we have defined so far. The computer responds with N and X.

3.5 How to Calculate a Mean

We have now defined by assignment two variable names, X and N: an array X and the number of elements, N, in X. This is all that we need to calculate a mean. The calculation is easy. Key in

```
      (+/X)÷N
¯1.7143
```

and we have indeed obtained the mean. But how? Let us try this again. Key in

```
      Y←2 4 6 8 4 2 6
      (+/Y)÷N
4.5714
```

Reduction /

Apparently the symbols +/, when applied to an array, add up the elements of the array. Mathematically, for an N-element array X this is $X_1 + X_2 + X_3 + \cdots + X_N$, or, more compactly, $\sum_{i=1}^{N} X_i$. The symbol / represents an operation on arrays called reduction, and reduction can be used with a large number of mathematical functions including +, -, ×, and ÷. Let f represent one of the arithmetic functions. Then (f/array) tells the computer to insert the function f between each element of the array and then perform all functions, but remember that it does so from *right to left!* Thus +/Y produces (i.e., is equivalent to) 2 + 4 + 6 + 8 + 4 + 2 + 6.

As another example, suppose L is a variable name of an array with three elements which represent the dimensions of a box, and you want to calculate the volume of the box. In APL, this problem is solved by typing ×/L. For example:

```
      L←3 2 5
      ×/L
30
```

Shape ρ

Let us return to calculating the mean. It would be most convenient if we did not have to count the number of elements in an array. Why not have the computer do it? Why not indeed! For this we use a little symbol called the shape function, ρ, which is the upper shift R key. Let's try it. Type

```
      ρX
7
      ρ(1&2&3&4)
4
      ρ1&2&3&4
4
```

So the argument of the function shape, ρ, can be a variable name or an array, and the result is the number of elements in the array. What about the shape (length) of the variable *N*, which is a scalar? Typing ρ*N*, for example, produces no response since a scalar has no dimension associated with it. As we will see, a scalar and an array with one element in it are different animals.

Arithmetic Mean

When we calculated the mean of the array *X*, we remembered that the computer reads APL statements from right to left, so that writing (+/*X*)÷*N* meant that the elements of *X* were added together and then divided by *N*. What would have happened if we had written +/*X*÷*N*? Each element of *X* would have been divided by *N*, and then the array of results summed. Both mathematical procedures theoretically give the same answers, but the second method is both slower and less computationally accurate if *N* is very large.

Let us review what we have learned about computing the mean of an array of numbers. Suppose you are given the array *X*. That is, *X* is in the computer ready for you to use, but you know nothing else about it. Problem: calculate the mean and find out how many elements there are in *X*. Here is one solution:

Monadic ρ

```
      N←ρX
      M←(+/X)÷N
      N
7
      M
¯1.7143
```

One thing to notice about the above is that:

(a) if you perform a function and assign the result to a variable, the result is stored under the variable's name and nothing is printed or displayed until you execute the name of the variable;

(b) if you perform a function and do not store the result, it will be displayed immediately;

(c) the values assigned by you to N and M will remain in the computer until you log-off or you redefine the variable name. For example:

```
      N
7
      N←9
      N
9
```

Do you remember all the variables you have defined? Type in the system command `)VARS` and see if you are right—the computer knows!

What should we do if we would like an array of the partial sums (sometimes called running totals) or partial means of X? That is, suppose we want the array

```
      1 (1+2) & (1+2+¯3) & (1+2+¯3+¯4) & (1+2+¯3+¯4+5)... &

      (1+2+¯3+¯4+5+¯6+¯7)
```

*Scan *

This is obtained by the symbols plus scan: \. The operation *scan* works in a manner similar to reduction except that after inserting the function "f" between each element of the array, the first element is kept, then the first pair of elements are reduced from right to left, then the first three, and so on. Try

```
      +\X
1 3 0 ¯4 1 ¯5 ¯12
      +\Y
2 6 12 20 24 26 32
```

3.6 Two Other Measures of Central Tendency: The Geometric and Harmonic Means

Geometric Mean

The geometric mean of N values is the Nth root of their product. Mathematically, one has

$$g = (x_1 \times x_2 \times x_3 \cdots \times x_N)^{1/N}$$

or

$$g = \left(\prod_{i=1}^{N} x_i \right)^{1/N}$$

How might we get the computer to calculate the geometric means of the arrays X and Y defined above? We have to learn some new functions first.

Raising a number to a power, taking logs, and related functions are computed as shown in Table 1. The mathematical function is given on the left, the corresponding general computer programming statement is given in the middle, examples are shown on the right, and the keying of the symbols is shown below the table.

Logarithm ⍟ and Exponential ⋆ Functions

Note that both ⋆ and ⍟ can be used as monadic (single argument) or dyadic (two argument) functions. The first and third rows show the dyadic uses and the second and fourth rows the monadic ones.

Table 1

Mathematical Statement	APL Statement	M/D	Examples	
	Exponential and Logarithmic Functions			
A^B	$A \star B$	D	5⋆2 25	3.2⋆0.6 2.0095
e^B ($e \cong 2.7183$)	$\star B$	M	⋆1 2.7183	⋆0.032 1.0325
$\log_B A$ (log of A to base B)	B ⍟ A	D	10⍟1 0	2⍟8 3
$\log_e A$ (or $\ln A$)	⍟ A	M	⍟1 0	⍟3.2 1.1632

M is Monadic, D is Dyadic.
⋆ is typed as upper shift *P* key.
⍟ is typed as upper shift *P* key, backspace, and upper shift *O*, to the left of *P*, *not* the zero key.

Logarithm and Exponential Functions ⍟ ⋆

⋆ and ⍟ are inverse functions of each other. For example,

```
      ⋆⍟3
3
      3⍟3⋆2
2
```

With the above functions we can now compute the geometric mean of an array of numbers. The geometric mean for the array *DATA* is:

```
      DATA←1.1 1.2 1.3 1.4 1.5 1.6 1.7
      N
7
      G←(×/DATA)⋆1÷N
      G
1.3855
```

and

```
      G←(×/Y)⋆÷N
      G
4.0679
```

In the former example, multiplicative reduction on `DATA` yields a result equivalent to the mathematical statement $\Pi_{i=1}^{N} D_i$, where D_i is the ith element in the array `DATA`. The remainder of the expression produces the Nth root of the product.

The second example illustrates a practical use of the monadic function ÷ that we discussed earlier, namely the inverse or reciprocal.

Harmonic Mean

The harmonic mean is the reciprocal of the arithmetic mean of the reciprocals. Mathematically,

$$h = N \Big/ \left(\sum_{i=1}^{N} (1/x_i) \right)$$

In APL, this is simply

```
      H←N÷(+/÷X)
      H
8.6726
```

and

```
      H←N÷(+/÷Y)
      H
3.5745
```

3.7 Sample Variance and Standard Deviation

Sample Variance

Calculating means presents us with few difficulties. What about calculating the variance and its square root, the standard deviation? The mathematical formula for the sample variance is simple enough.

$$s^2 = \sum_{i=1}^{N} (x_i - \bar{x})^2/(N - 1),$$

where N is the number of observations N_i and $\bar{x}$ is their arithmetic mean.

If we know $\bar{x}$, the solution is apparent. Consider the following APL expression, which is a series of functions linked together to make up the APL equivalent of a mathematical expression. (DO NOT TYPE IT IN YET!)

```
      (+/((X-M)*2))÷N-1
```

Parentheses

The above expression was obtained by the following line of thought. Let M represent the arithmetic mean $\bar{x}$. Then the expression $\Sigma(x_i - \bar{x})^2$ is in APL `+/(X-M)*2`; the term inside the parentheses is an array $x_1 - \bar{x}, x_2 - \bar{x}, \ldots, x_N - \bar{x}$, each term of which is squared, and then plus reduction is performed on the resulting array. Remember that the computer reads APL statements from right to left, and expressions in parentheses are evaluated as soon as they are encountered by the computer. In the above expression the array `(X-M)` is calculated first, then each element is squared. With a number of pairs of parentheses embedded in each other as above, the expression within the *innermost* parentheses is evaluated *first*, then the expression within the next outside ones, and so on. The resulting array is plus reduced (i.e., the elements of the array are added), and finally, the summed array is divided by (N-1).

Now we are ready to try out our expression. First type

```
      )VARS
G       H       L       M       N       X       Y
```

just to make sure we still have N and X stored in the computer. If you do not get N and X listed, then you probably signed off after you last used those variables. If that is the case, enter them into the computer again (X is given on page 18 and N is obtained by N←ρX). Now type

Sample Standard Deviation

```
      M←(+/X)÷N
      V←(+/((X-M)*2))÷N-1
      SD←V*.5
      M
¯1.7143
      V
19.905
      SD
4.4615
```

If you did not get the same results, check first to see if the mean value is the same. If it is not, your X array may not match that shown on page 18, or the value of N may be incorrect. If V is wrong while M is right, check your APL expression very carefully to make sure that it is exactly like the one shown above.

Checking Parentheses

One little hint about keeping parentheses properly paired up: going from right to left, add 1 each time you hit a right parenthesis and subtract 1 each time you hit a left parenthesis; when you are out of parentheses, the answer should be zero, because the number of right and left parentheses should be equal. If they are not, find the missing or extra parenthesis. For example,

```
      V←(+/((X-M)*2)÷N-1
        ↑  ↑↑    ↑  ↑
       -1  01    2  1
```

The count ends at -1, so we have either a missing right parenthesis or an extra left parenthesis. To find out which, go to the innermost pair of parentheses and work outwards in both directions. Thus

(X-M)	looks alright
((X-M)*2)	looks alright

```
(+/((X-M)*2)    here is the error, a
            ↑   missing right parenthesis
```

If instead we were to delete the first left parenthesis, we would get the "right answer," but in a very inefficient manner. In the latter case the squared elements of the array `(X-M)` would each be divided by `(N-1)` and the quotients added. In the original expression, the squared elements are added and then the sum is divided by `(N-1)` *once*.

3.8 Correcting Typing Errors

In keying the above APL expressions you may have made some typing errors—a common error is to have too many or too few parentheses. So far it has been easy enough to hit RETURN, get some error message, and redo the expression. However, you can see that as your APL expressions get longer, this will become a nuisance, so let's see how to correct a line *while* it is being typed, that is, before hitting the EXECUTE key. Backspace until the cursor on the terminal head (a little device that indicates where the next character will be typed) is at the beginning of your first mistake (i.e., everything to the *left* of the cursor is correct), then hit the ATTN (Attention) key. Now type the remaining part of the line. Alternatively, hit the "line feed" key on the right-hand side of the terminal. For example, suppose that you are working at a "hard copy" terminal (that is, one that prints on paper), that you have typed

```
V← (+/ ((X-M)*2÷N-1
```

and that you realize your error before hitting the EXECUTE key. Backspace to the division sign, hit "line feed," which advances the paper one row (and tells the computer to add the new characters to the previous line), and then complete the line correctly. You will have

```
V← (+/((X-M)*2)÷N-1
               )÷N-1
```

and the computer will correct your error as soon as you hit EXECUTE.

Editing lines on the IBM 5120 and many other CRT* terminals is even easier. You can simply backspace and type in the correct characters. Some terminals have the ability to insert characters within a line. You space back until you reach the last correct character, hold down a special key ("command" on the 5120 series), and press the right arrow on the top row of keys.** The result is

```
V←(+/((X-M)*2)&÷N-1
```

* Cathode Ray Tube—electronics jargon for a television screen.

** This is true on terminals that have an addressable cursor. For others, the correction process is more elaborate. In some cases, each character may need to be erased. In others it may be easier to just replace the whole line.

In effect, you moved the four characters `÷N-1` one space to the right and held the cursor at its original point. You now type the missing ")":

```
V←(+/((X-M)*2))÷N-1
```

The procedure used for editing lines is specific to the computer system you are using and also to the particular terminal interfaced to that system. CRT's generally provide the most flexibility, but having a written or hard copy of your session is often extremely valuable. You will have to consult the computer center personnel for specific editing procedures, as such procedures are not explicitly part of the APL language.

If we refer again to the APL statements on page 24, we notice that the three lines of statements that calculate M, V, and SD must be executed in precisely the order shown. This is because the second line needs the result of the first, and the third needs the result of the second. We are beginning to discover that we will have to develop tools more powerful than those that we have used so far. This will be the subject of the next chapter. Meanwhile, we will conclude this chapter with a way of calculating sample means (arithmetic) and variances from sample probabilities of success (see, for example, Kmenta, Chapter 2).

3.9 Mean and Variance of Sample Probabilities

Suppose we are interested in estimating the probability of getting a seven when we roll a pair of dice. (Of course, it is easy to see that if we have clean, unloaded dice, the probability is $1/6$, but we might want to check our dice.) One way to do this experimentally is to roll a pair of dice N times and then divide the number of successes (number of times you got a seven) by N. But this is merely an estimate. How might we estimate the mean and variance of this estimate? One way would be to repeat the above experiment a large number of times, say NN, and then to calculate the mean and variance of the estimates of the probability of a seven that were obtained in each trial.

Suppose you have data obtained from $NN = 100$ replications of a dice tossing experiment in which $N = 4$ tosses were made. In any one experiment of four tosses you could obtain zero to four sevens—five possibilities in all. The *estimated* probability from each experiment could vary from zero (equal to 0 successes divided by 4, the number of trials), to 1 (4 successes in 4 trials). As we suspect, if our dice are unloaded, PR, the probability of getting a seven is $1/6, PR = 0.166\ldots$. From each experiment we get an estimate, say $\widehat{PR}$, which can take one of 5 discrete values, viz., 0, 0.25, 0.5, 0.75, and 1.0. Let $\widehat{PR}_1 = 0$, $\widehat{PR}_2 = 0.25, \ldots, \widehat{PR}_5 = 1.0$. If NN equals 100, then we can count the number of times n_i that we get each estimate $\widehat{PR}_i$, $i = 1, 2, \ldots, 5$, in 100 trials. These five numbers $n_1, n_2, \ldots, n_5$, whose sum is 100, are called *absolute* frequencies. If we divide each n_i by 100 we get five *relative* frequencies whose sum is 1.0. Let's call the relative frequencies fr_i, $i = 1, 2, \ldots, 5$. fr_i is merely the proportion of the NN repetitions of our experiment that yielded PR_i as the estimated probability. That is, $fr_i = n_i/NN$.

Mean & Variance of Sample Probabilities

The first thing that we must determine is the mean estimate $\widehat{PR}$. The mathematical statement of the answer is simple: $MF = \Sigma_{i=1}^{5} fr_i \widehat{PR}_i$, where MF represents the mean of the sample probabilities obtained from the observed relative frequencies fr_i. (MF is a sample mean of sampling proportions $\widehat{PR}_i, i = 1, 2, \ldots, 5$.) The variance VF is given by the expression, $VF = \Sigma_{i=1}^{5} fr_i(\widehat{PR}_i - MF)^2$.

The calculation of MF and VF in APL, although straightforward, introduces us to yet another function. Let us suppose that we have the following data from a sampling experiment in which $NN = 100$: $fr_1 = 0.01$, $fr_2 = 0.06, fr_3 = 0.28, fr_4 = 0.42, fr_5 = 0.23$. Enter the APL statements

```
FR←0.01 0.06 0.28 0.42 0.23

PR←0.0 0.25 0.50 0.75 1.0
```

We now have all the data we need ready and waiting in the two arrays FR and PR.

Inner Product
+ . ×

To get what we want requires an operation called the "inner product," and the version we want here is typed by keying plus, period or decimal point, and then multiply. The expression for MF then is simply FR+.×PR. The APL code tells the computer to take in turn each element of the array FR, multiply it by the corresponding element in PR, and add the products. The calculation of the sample mean and variance may thus be carried out by typing

```
      MF←FR+.×PR

      VF←FR+.× (PR-MF)*2

      MF
0.7

      VF
0.05
```

From our results it would seem that our dice are definitely loaded!

Now that you have learned how to calculate some basic statistics, you will be anxious to try your hand at some more realistic data. If you try to enter somewhat more data than we have been using so far, you will run into a little problem. The problem is that the computer will limit the amount of data you can enter. The limit is usually 80, 128, or 160 characters. You get to the right-hand end of a line and you either cannot enter more data, or the new data replaces your previous entry.

Line Continuation with ,⎕
Entering Data on Two Lines

The solution to this problem is not difficult. Whenever you want to continue entering data on the next line, simply close off the current line by the symbols ,⎕ and hit RETURN (or EXECUTE). The symbols are a comma followed by an upper case L. ⎕ is called "quad," and you will be meeting this useful operator again. The computer will respond on the next line by printing ⎕: after which you carry on entering data until finished. You can use this device to enter as much data as you wish.

Summary

The Computer Reads from Right to Left.
Expressions in Innermost Pairs of Parentheses Are Executed First.

Arrays of numbers and scalars can be given names by *assignment,* e.g., `X←1 2 3` assigns the name X to the array 1 2 3. There are name limitations on the symbols that can be used in making up a name. Names should begin with a letter—after that any other alphabetic character or number is okay. The name can be as long as you want (only the first 77 characters will be recognized); don't put blanks inside the name.

Dyadic functions have two arguments.

Monadic functions have one argument; the order is "*f* array" or "*f* number."

Many symbols, such as +, -, ×, ÷, do double duty and represent both dyadic functions as well as monadic ones.

System Commands are instructions *about* the operation of the computer. They are indicated by a right parenthesis:).

`)VARS` is a system command which instructs the computer to list all the variable names assigned by the user.

Reduction, /, a monadic function, is used with the arithmetic functions on arrays:

$$f/x_1, x_2, \ldots, x_n \quad \text{produces} \quad x_1 f x_2 f x_3 f \ldots f x_n$$

Shape, ρ, displays the number of elements in an array.

Arithmetic Mean (of an array X): Mathematically, $M = (\Sigma_{i=1}^{N} x_i)/N$, and in APL

```
N← ρX

M← (+/X)÷N
```

Scan, \, a monadic function, is used with the arithmetic functions on arrays:
$f\backslash x_1, x_2, \ldots, x_n$ produces $Y_1, Y_2, \ldots, Y_n$ where $Y_1 = x_1$, $Y_2 = x_1 f x_2$, $Y_3 = x_1 f x_2 f x_3, \ldots, Y_n = x_1 f x_2 f x_3, \ldots, f x_n$.

The exponential function, ⋆, has a monadic and a dyadic use; see Table 1.

The logarithmic function, ⍟, has a monadic and a dyadic use; see Table 1.

Geometric mean (of an array X): Mathematically, $G = (\Pi_{i=1}^{N}(x_i))^{1/N}$. In APL,

```
N← ρ X

G← ( ×/X)⋆÷N
```

Harmonic Mean (of an array X): Mathematically, $H = N(\Sigma_{i=1}^{N} x_i^{-1})^{-1}$, and in APL,

```
H←N÷+/÷X
```

Sample Variance (of an array X): Mathematically, $V = \Sigma_{i=1}^{N} (x_i - M)^2 / (N - 1)$, where M is the arithmetic mean. In APL,

```
V← ( +/ (( X-M)*2))÷N-1
```

The Sample Standard Deviation (of an array X) is the square root of the sample variance.

Inner product (between two arrays X and Y of equal length): Mathematically, $IP = x_1y_1 + x_2y_2 + \cdot \cdot \cdot + x_ny_n$. In APL,

```
IP←X+.×Y
```

Mean (arithmetic) for Sample Proportions: Mathematically, $MF = \Sigma_{i=1}^{K} fr_i\widehat{PR}_i$, where K is the number of cells and fr_i is the relative frequency of the ith value of $\widehat{PR}_i$, the ith sample proportion. In APL,

```
MF←FR+.×PR
```

Variance for Sample Proportions: Mathematically, $VF = \Sigma_{i=1}^{K} fr_i(\widehat{PR}_i - MF)^2$. In APL,

```
VF←FR+.× (PR-M)*2
```

How to continue entering data over more than one or two lines: use ,⎕ at the end of a line of input on the terminal.

Exercises

Chapter 3

APL Practice

1. Let's explore the uses of the functions defined in this chapter.
   Let `P←3` and `Y←1 2 3 4`.
   - (a) `ρ P`
   - (b) `ρ 3`
   - (c) `ρρ 3`
   - (d) `ρρ P`
   - (e) `ρ Y`
   - (f) `ρρ Y`
   - (g) `+/Y` and compare it carefully with `+\Y`
   - (h) `(+/Y⍟Y)` which is the same as `ρY`. Why?

2. Assign the values `1 2 3 4` to `X`, the values `¯1 ¯2 ¯3 ¯4` to `W`, and define `Z` by `Z←*X`.
   - (a) `1*X`
   - (b) `X*1`
   - (c) `2*X`
   - (d) `X*2`
   - (e) `0*X`
   - (f) `X⍟0`
   - (g) `7*W`
   - (h) `X*¯1`
   - (i) `1⍟X`
   - (j) `X⍟1`
   - (k) `2⍟X`
   - (l) `X⍟2`
   - (m) `X*0`
   - (n) `0⍟W`
   - (o) `¯1⍟X`
   - (p) `X⍟¯1`
   - (q) `⍟Z` which is `X`
   - (r) `+/X÷ ρ X`
   - (s) `( +/X)÷ ρ X`
   - (t) What is the difference between (r) and (s)?
   - (u) `+/X×W*2` which is the same as `X+.×W*2`.
   - (v) `-\X` and compare with `-/X`.
   - (w) `-\W` and compare with `-/W`.

3. Evaluate the polynomials $f(x) = x^2 - 5x + 6$, $f(x) = (x^5 - 1)(x^2 - 2)$, and $f(x) = x^4 - 3x^2 + 2$ for `X←¯5&¯3&¯1&0&1&2&3`.

4. Practice the right to left rule by solving:

   (a) `10+¯3+-5` (f) `-/1&2&3÷3`

   (b) `-3&¯5-10` (g) `-/÷1-3&-¯2`

   (c) `÷3&-5&-10` (h) `+\2¯34`

   (d) `3-X÷6-10` (i) `+\10*123`

   (e) `+/÷1&2&3`

   You should be able to find the answers without using the terminal and then use the terminal to check.

5. Let the arrays `X` and `Y` be defined by

   `X←1&2&3&4&...&10, Y←¯2+X*.3.`

   Practice the algebra of summations by trying:

	APL Form	Math Form	Explanation
(a)	`+/÷X`	$\Sigma_{i=1}^{N} 1/X_i$	the sum of the reciprocals of the elements of X
(b)	`+/X-1`	$\Sigma_{i=1}^{N} (X_i - 1)$	the sum of the differences $X_i - 1$
(c)	`+/X*2`	$\Sigma_{i=1}^{N} X_i^2$	the sum of the squares of the elements of X
(d)	`(+/X)*2`	$(\Sigma_{i=1}^{N} X_i)^2$	the squared sum of the X_i
(e)	`+/⍟X`	$\Sigma_{i=1}^{N} \ln X_i$	the sum of the natural log of the X_i (natural log is log to base e)
(f)	`+/*X`	$\Sigma_{i=1}^{N} e^{X_i}$	the sum of e raised to the powers X_i
(g)	`-/X`	$\Sigma_{i=1}^{N} (-1)^{i-1} X_i$	
(h)	`-/-X`	$\Sigma_{i=1}^{N} (-1)^{i} X_i$	
(i)	`+/X×Y`	$\Sigma_{i=1}^{N} X_i Y_i$	These sums occur very frequently in
(j)	`+/Y×X*2`	$\Sigma_{i=1}^{N} X_i^2 Y_i$	regression analysis; see Chapter 8.

6. Verify that the first element of `X` is equal to the first element of `+\X`, while the last element of `+\X` is equal to `+/X`.

7. Here are some more summation formulae. Try expressing each in terms of APL functions.

   (a) $\Sigma_{i=1}^{N} X_i - 1$
   (b) $\Sigma_{i=1}^{N} e^{x2_i}$
   (c) $\Sigma_{i=1}^{N} (X_i + 3)^2$
   (d) $\Sigma_{i=1}^{N} (-1)^i (1/X_i)$
   (e) $(\Sigma_{i=1}^{N} X_i)/N$
   (f) $1/\Sigma_{i=1}^{N} (1/X_i)$
   (g) $\Sigma_{i=1}^{N} (X_i^2 - 5)^2$
   (h) Show that $\Sigma_{i=1}^{N} kX_i = k\Sigma_{i=1}^{N} X_i$
   (i) Show that $\Sigma_{i=1}^{N} X_i^2 + 4\,\Sigma_{i=1}^{N} X_i + N^2 = \Sigma_{i=1}^{N} X_i^2 + 2N\Sigma_{i=1}^{N} X_i + N2^2$
   (j) Show that $\Sigma_{i=1}^{N} X_i^2 - (\Sigma_{i=1}^{N} X_i)^2/N$
   $= \Sigma_{i=1}^{N} X_i^2 - 1/N(\Sigma_{i=1}^{N} X_i)^2$

where N is the number of elements of X. As there are several ways to write each of the above, the suggested procedures in the solutions may not always be the same as yours. You can, however, check that your procedure is correct by computing the numerical solutions. Use the list X defined in exercise number 5.

8. Interpret the following APL statements in mathematical form:

(a) `+/÷X` (d) `÷+/÷X×Y`

(b) `-/÷X` (e) `÷+/ ( X-+/X÷ ρ X)*2`

(c) `-/÷-X` (f) `( X×Y)- ( +/X)× ( +/Y)`

9. Which of the following are invalid names?

(a) `ABE LINCOLN`

(b) `B*`

(c) `LOUIS̥ THE 14TH`

(d) `XXX+YYY`

(e) `X1X2X3`

(f) `2W+3Y`

(g) `IWILLBEHERETOMORROWATTHREE`

Statistical Applications

1. When the elements of an array are ratios, the geometric mean may be a more useful measure of central location than the arithmetic mean would be. If an array has elements that are rates of change, then the harmonic mean is usually preferable.

Consider the following data:

Year	U.S. Total Residential Debt Outstanding in Billions of Dollars
1966	274.2
1967	292.0
1968	312.8
1969	335.9
1970	357.8
1971	398.0
1972	454.5
1973	509.8
1974	549.8
1975	593.0
1976	655.0
1977	711.2 (estimated)

Source: U.S. League of Savings Association Publications #24, 1977, p. 28.

(a) Find the arithmetic mean of U.S. residential debt outstanding during the 12 years.

(b) Find the ratio of each year's debt to the previous year's.

(c) What is the geometric mean of these ratios?

(d) Find the percentage increase of the debt series from each year to the next. (This is equal to the ratios in (b) minus 1.)

(e) What is the harmonic mean of the percentage increases?

(f) What is the mean rate of growth in U.S. residential debt?

(g) Explain intuitively your answers to (a), (c), and (f).

2. In the following table CE represents the number of cracked eggs in a carton (each carton contains 12 eggs). CN is the number of cartons out of a sample of 60 cartons, randomly selected from a shipment of 2,000, that contain cracked eggs.

CE	0	1	2	3	4	5	6	7	8	9	10	11	12
CN	0	5	7	11	16	8	4	5	2	1	0	0	1

(a) What is the average number of cracked eggs in each carton?

(b) What percentage of cartons have fewer than 2 cracked eggs?

(c) On the basis of your answers to (a) and (b), should the shipment be accepted? (A shipment is acceptable when 8% or less of the eggs are cracked.)

3. Let $W_i, i = 1, \ldots, 5$, take the values 3, 4, 7, 5, 11, and $Y_i = 3 + 2W_i$. Find $\bar{W}$, $\bar{Y}$, $S_W^2 = (\Sigma_{i=1}^5 (W_i - \bar{W})^2)/4$, $S_Y^2 = (\Sigma_{i=1}^5 (Y_i - \bar{Y})^2)/4$, $S_W = \sqrt{S_W^2}$, $S_Y = \sqrt{S_Y^2}$. Verify that $\bar{Y} = 3 + 2\bar{W}$, $S_Y^2 = 4S_W^2$, and $S_Y = 2S_W$.

4. Let r represent a list of estimates of the interest rate for next year: 5%, 6%, 7%, 8%, 9%, 10%. Suppose we believe that the respective probabilities that these values will occur are 0.1, 0.2, 0.3, 0.2, 0.1, 0.1. Find the expected value of the interest rate. What is the probability that the interest rate will neither fall below 6% nor exceed 9%? (*Note:* the expected value of a variable X which can take on only discrete values is defined by $\Sigma_{i=1}^N X_i P_i$, where P_i is the probability that X_i will occur.)

5. Consider a gamble wherein a fair coin is repeatedly tossed until a head turns up. If a head is obtained on the first toss the payoff is $2; it is $4 if a head is obtained on the second toss, $8 on the third toss, and so on. Use the computer to find the expected return if the coin is tossed at most a total of one hundred times. (A new wager is made after each time that a head appears.)

6. The following measurements (in dollars rounded to the nearest integer) represent the increase (if positive) or decrease (if negative) in the daily closing price of General Motors and General Electric common stocks for 10 consecutive days.

GM	2	$^-1$	3	0	0	1	$^-1$	2	4	0
GE	3	4	$^-1$	0	$^-1$	$^-2$	0	4	3	$^-2$

(a) Which stock would you prefer with respect to average daily return?

(b) If the larger the variance of a security the larger the risk, which is the riskier security?

7. Imagine yourself sitting at the roulette table in Las Vegas, having lost the family fortune, and being left with only \$200 at your disposal. The roulette wheel will be spun about seventy more times before the table is closed. There are 36 numbers and a double zero on the wheel. You decide to bet \$3 on the number 11 repeatedly. The rule of the game is as follows: If 11 comes up you get \$105 (\$3 × 35), the payoff is \$35 for a \$1 bet; if any other number comes up you lose the \$3. What is your expected cash position at the end of the night?

4

How to Write Your Own Function

In this chapter, after defining a few more APL expressions, we introduce the important concept of a function which you can write yourself. At the end of this chapter your ability to apply APL will have taken a big leap forward. Let's press on.

4.1 The Sample Median

One measure of central tendency that is in widespread use is the median. The median is that value for which half of the sample values are less than or equal to it and half are greater than or equal to it. The median for n observations is defined mathematically by

Sample Median

$$M = \begin{cases} (x_{n/2} + x_{n/2+1})/2 & \text{if } n \text{ is even} \\ x_k,\ k = (n-1)/2 + 1 & \text{if } n \text{ is odd} \end{cases} \quad (4.1)$$

where $x_1 \leq x_2 \leq \ldots \leq x_n$ are the order statistics obtained from n observations. That is, the n observations are reordered so that the smallest observation is first (x_1) and the largest is last (x_n).

Calculating M in APL might seem to be a formidable task, but in fact it is quite simple. In addition, calculating the median will introduce us to some useful programming tools.

We have two problems to solve—discovering whether n is even or not, and reordering the array to get the order statistics. That is, we want an array with the smallest observed number first, the largest last, and with each number less than or equal to the number on its right. Let us begin with some useful new APL tools.

The Residue Function

Residue, |

The first of these tools is the residue function, |, which is upper shift M. The residue function applied to two numbers A and B, denoted by A|B, yields the A residue of B, and is the "remainder" left over after dividing A into B. For example, the 2 residue of 3 is 1, the 3 residue of 8 is 2, the 4 residue of 8 is 0, and so on. Of particular interest is the 1 residue of a positive decimal number; thus

```
      1|2.5&0.6&4.0
0.5 0.6 0
```

In other words, the 1 residue of a positive decimal number is simply the decimal fraction of that number. Let's try some more examples:

```
      3| 0 1 2 3 4 5 6 7 8 9 10
0 1 2 0 1 2 0 1 2 0 1
      10| 11&102&1032&11021
1 2 2 1
      1| ¯6.0&¯6.4&¯0.3
0 0.6 0.7
      2|3&4.0&¯3.0&6.2&¯6.2&¯4.0
1 0 1 0.2 1.8 0
      3|4.0&¯4.0&5.0&¯5.0&4.2&¯4.2&¯4.8
1 2 2 1 1.2 1.8 1.2
```

The results for the first two arrays are clear enough, but what about the others? Close inspection reveals a difficulty in interpretation of the results only when we try to get the residue of a negative number. What is occurring with negative numbers will become clear as soon as we understand the residue operation with positive numbers.

If we look back at the first example we notice a recurring sequence 0 1 2. Suppose that we extend the array to include negative numbers, say

```
      3| ¯10 ¯9 ¯8 ¯7 ¯6 ¯5 ¯4 ¯3 ¯2 ¯1 0 1 2 3 4 5 6 7 8 9 10
2 0 1 2 0 1 2 0 1 2 0 1 2 0 1 2 0 1 2 0 1
```

Note that the pattern is exactly the same.

Absolute Value, |

However, 3|4 is 1 and 3|¯4 is 2, so the result is not simply the residue of the absolute value of the righthand list. In fact the monadic use of the | symbol *is* the absolute value function. For example

```
      |¯4
4
      |¯2.1 ¯1 0 1.2 3.1
2.1&1&0&1.2&3.1
```

Absolute Value and Residue to Compute Fractional Parts of Positive & Negative Numbers

So we could find the fractional parts of elements of a vector that had both positive and negative elements by

```
      1||¯2.1 ¯1 0 1.2 3.1
0.1 0 0 0.2 0.1
```

Remember, we proceed from right to left, first finding the absolute values of the array by using the monadic | function and then using the dyadic | function to find the fractional parts.

Logic of Residue Function

While we have shown that the residue is consistent in its operation on positive and negative numbers, we need to explain the logic behind this consistency. Here is one way to think about it. The residue for a positive righthand argument is obtained by successively subtracting the lefthand value until the result is less than the lefthand value. If we have 3|7 we subtract 3 from 7, yielding 4, then subtract 3 from 4, leaving 1, which is less than 3 so we stop. The last value is the residue. When the righthand value is *negative* we *add* the lefthand value to the right until the result is positive. So if we have 3|¯7, we add 3 to ¯7, yielding ¯4, then add 3 to ¯4, leaving ¯1, then add 3 to ¯1 to get 2. This process defines the residue function for positive lefthand and negative righthand values.

Finally, here is a diagram of the results of the first example:

Statement 3|	¯10	¯9	¯8	¯7	¯6	¯5	¯4	¯3	¯2	¯1	0	1	2	3	4	5	6	7	8	9	10
Result	2	0	1	2	0	1	2	0	1	2	0	1	2	0	1	2	0	1	2	0	1
Index	1	2	3	4	5	6	7	8	9	10	11	12	13	14	15	16	17	18	19	20	21

Diagram 1 Illustrating the Value of the Residue When Dividing Numbers by Three*

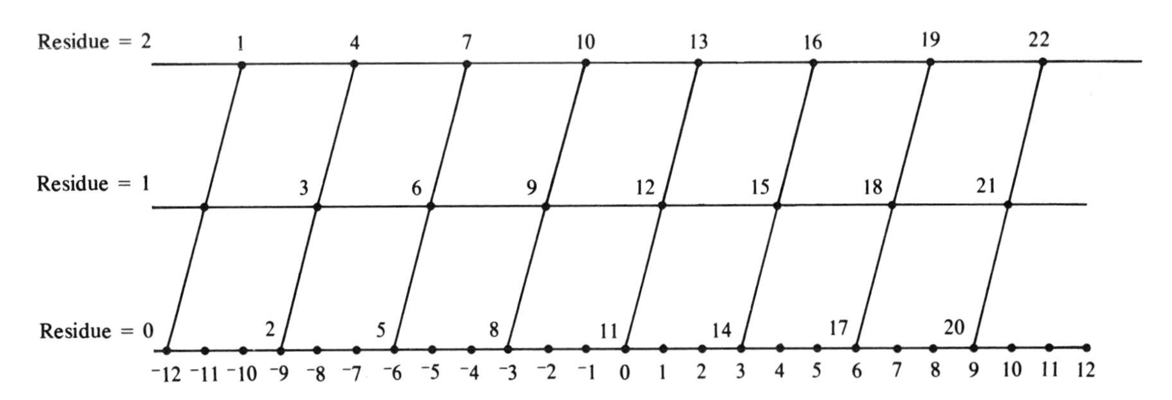

* Right argument of residue on x axis, residue on y axis, numbered points are the index values (see above). Locate 4 on the bottom of the diagram, move vertically until you cross the diagonal line, read the residual value 1 on the left. Check to see that it is element 15.

Parity of the Number of Elements in an Array

Arrays: Parity

After this lengthy digression we can take the first step in finding the median. Does the array have an even or odd number of elements? If you have re-logged-on to do this chapter, you will have lost the *X* and *Y* arrays defined in Chapter 3. If so, please enter these values first:

```
X←1&2&¯3&¯4&5&¯6&¯7
Y←2&4&6&8&4&2&6
```

Solution:

```
N← ρ X
2|N
```

If the result is 0 we have an even number of elements (*N* is divisible by 2 with no remainder), and if the result is 1 we have an odd number. In our example *N* is 7, so we have an odd number and the result of 2|*N* is 1. You may be wondering why we bother with all of this, since printing *N* tells us immediately if *N* is even or odd. The reason is that we are laying the foundation for the computer to make the decision itself, without our intervention.

Ordering the Elements of an Array

Ordering Arrays

The next problem is to reorder the array *X* so that the smallest element is first and the largest is last. To be safe and not lose the original order, let us store the reordered array *X* in a new array called *R*. The process of reordering an array introduces two new concepts, the grade up (or down) function and the simple mysteries of indexing. Let us first consider the indexing of arrays and the indexing function.

Indexing Arrays

Since a one-dimensional array is merely a list of elements, it seems natural to give the position of each element in the list an index value and to let that value be the position of the element in the list; in short, the first element *going from left to right* is element number 1, the second is number 2, and so on. APL lets us refer to the elements in an array in a simple manner. Type out the following:

```
      W←9&10&3&8&12&0&1&5
      W
9 10 3 8 12 0 1 5
      W[1]
9
      W[5]
12
      W[1&2&3]
9 10 3
      W[3&1&2]
```

```
3 9 10
      W
9 10 3 8 12 0 1 5
```

The examples illustrate what is happening. The indexing of an array allows you to select from an array the individual elements of an array in any order. In addition, specific elements in an array can be replaced very easily; for example,

```
      W[1&4]
9 8
      W[1&4]←6&15
      W
6 10 3 15 12 0 1 5
```

The contents of [ ] can be any valid APL expression, provided only that it results in an array of integers, none of which is larger than ⍴*W*, the number of elements in the array *W*. Try

INDEX ERROR

```
      W[0]
      INDEX ERROR
      W[0]
       ^
      W[2 2 2]
10 10 10
      W[9]
      INDEX ERROR
      W[9]
       ^
      W[¯2]
      INDEX ERROR
      W[¯2]
       ^
```

Grade Up ⍋
Grade Down ⍒

Now consider the functions grade up, ⍋, and grade down, ⍒. Each is formed by typing one character on top of another (the | we used for residue (upper shift *M*), backspace, and then upper shift *H* for ∆ (delta) or *G* for ∇ (del)). Let's try it on the array *W* that we defined above:

```
      ⍋W
6 7 3 8 1 2 5 4
```

and

```
      ⍒W
4 5 2 1 8 3 7 6
```

To see what is happening here, write down the elements of W with their indices underneath them:

W:	6	10	3	15	12	0	1	5
index:	1	2	3	4	5	6	7	8

We see that ⍋ (grade up) gives us a new array in which the first element is the index value of the smallest element of W, the second element is the index value of the second smallest element of W, and so on. Grade down (⍒) also indicates the order of the array, except that the first index value in the new array is that of the largest element in W, the next is the index value of the second largest element of W, etc. The last index value in the new array is the position of the smallest element of W.

You might be able to guess what we should do now in order to get an array with the elements of W or X or Y listed in either ascending or descending order. Try

```
      WUP←W[⍋W]
      WUP
0 1 3 5 6 10 12 15
      WDOWN←W[⍒W]
      WDOWN
15 12 10 6 5 3 1 0
```

Sorting an Array

This is all you need to sort any list of numbers. The grade function orders the index values inside the square brackets, and then the elements represented by those positions are selected.

Calculation of the Sample Median

Sample Median

Now we can calculate the median. If our expression for discovering the parity of N tells us that N is odd, then by Eq. (4.1) the median value of the elements of X is element number $((N - 1)/2) + 1$ of the ordered array. (It is irrelevant here whether the elements of X are arranged in ascending or descending order.) Let's try it, but first let's recall what X and N are.

```
      X
1 2 ¯3 ¯4 5 ¯6 ¯7
      N
7
```

```
      XUP←X[⍋X]                ← rearranges X in ascending order
                                 and stores result in XUP

      M←XUP[1+ (N-1)÷2]        ← picks out the middle element
                                 of XUP
      M
¯3
```

If N is even, the median is:

```
      W
6 10 3 15 12 0 1 5
      WUP←W[⍋W]
      K← (ρ W)÷2
      M←WUP[K,K+1]             ← picks out the middle pair of
                                 elements
      M← ( +/M)÷2
                               ← averages them
      M
5.5
```

You can easily check by visual inspection that both calculations are correct. Each of these two sets of APL statements can be called a "routine"; a routine is a group of APL statements that perform some operation.

These two APL routines worked, but the third line in the second routine contains something strange—the contents of [ ] must be an *array* of integers, but what is the comma doing there? Why does WUP[K,K + 1] work? First, ask yourself what you would get if you typed in $K_{\&}K$ + 1, where K has the value 4. If you incorrectly read from left to right, you might believe that the answer would be 5 5. You would be incorrect in believing that K K + 1 is computed as "to the array $K_{\&}K$ add 1." APL expressions are computed from right to left. But we cannot separate K and K + 1 by parentheses alone, e.g., K (K + 1), since APL expects a function to appear between K and (K + 1). The function represented by "," is discussed next.

The Catenate Function

Catenate, ,

The solution to our little problem introduces a very useful concept and a useful function, the catenate function " ,", which is keyed in by typing a comma. The catenate function enables us to extend an array or to make arrays out of scalars. For example, let A and B denote arrays of length p and q, respectively, and let K and L be scalars. Then

A,B produces an array of length $(p + q)$ with the *A* elements coming first,
B,A produces an array of the same length, but the *B* elements come first,
A,K produces an array of length $(p + 1)$,
K,L produces an array of length 2,
,L produces an array of length 1.

So the solution to our problem is to create the array *K,K* + 1, a two-element array of which the first element is *K* and the second is *K* + 1.

We have now solved the problem of computing the median but, especially if *N* is even, it is a nuisance to type out all these statements each time we want the median. What would be convenient would be a function called, say, *MD*, such that typing $MD_{\&}X$ (where *X* is an array) produces the median of *X*. In short, we need a monadic function just like the ones provided by APL, like ⍋, ⍴, |, and so on. To meet this need, APL provides a method whereby the user can supplement the set of primitive APL functions (those available from the keyboard, such as ⋆, +, etc.) with user-defined functions. Once the user function is defined, it can be used over and over again just by calling it—for example, by typing $MD_{\&}X$, where *X* is an array, the median of which is defined by the function called *MD*.

4.2 Function Definition

Entering Function Definition Mode
∇

Our first step is to tell the computer that we want to define a function to be used later. This is done by typing ∇ (called "del"), which is the upper shift *G* key. We will also need to inform the computer when we are finished defining the new function. That is done by typing in ∇ again. Thus the computer expects that between the two ∇ symbols it will be receiving instructions that will define a new function.

Certain rules must be followed when defining functions, and we will consider these rules now. If we type ∇, which tells the computer "Function definition coming up," our next action must be to give the function a name. Once we have named our function, both we and the computer can refer to it even while we are in the process of defining it. But in APL we not only name a function, we can also give it some arguments—that is, something for the function to operate on.

Niladic Functions

Niladic Function

In §3.2 we defined monadic and dyadic functions—functions with one and two arguments, respectively. We can also have niladic functions, which do not have any arguments. A niladic function is called simply by typing ∇ and the name of the function and hitting the EXECUTE or RETURN key. It then carries out the procedure specified in the function—you do not give it any arguments to work on. Let's consider an example.

Suppose we want to run some experiments on the computer to check out the statistical properties of the probability estimator $\widehat{PR}$ discussed at the end of the last chapter. One of the first things we have to be able to do is to generate data similar to that which we would get from rolling a pair of dice. If *N*, the sample size, is 16 or 20, and *NN*, the number of repetitions of the experiment, is 100 or more, and if we want to see how the mean and variance change as we alter *N*, then we are facing a very large amount of dice throwing. Such an activity may not be your favorite way of spending a Sunday afternoon in the sun, but if it is pouring rain, go ahead and roll the dice. Fortunately instead of rolling dice we can use the computer to simulate the process of actually tossing them. So, let's generate some data.

We begin by defining a function that will give us one roll of a pair of dice. Before beginning to define our own functions, let us clear the decks, so to speak, of any other nonprimitive functions that might be already in our "workspace" and might give us some difficulties and strange responses.

)CLEAR

We do this by typing *)CLEAR*, and the computer responds by telling us that our workspace (you can think of it as the part of the computer that we are using) is clear. WS stands for workspace. You will recall that the symbol) indicates that a system command is coming up. Type

```
      )CLEAR
CLEAR WS
      ∇R←DICE
[1]
```

The use of ∇ in the first line is clear enough; it tells the computer that we are defining a function. In more technical language, it puts us into "function definition mode" instead of "execution mode." The word *DICE* is the name of our function, but why *R*? We will explain that one in a minute.

The computer responded with a [1]. This is its way of telling you that it understands that you are defining a function, and [1] is the number of the first statement for you to write. The computer now expects you to write an APL statement, so let's oblige.

Roll ?

We will be using a special APL symbol, *?*, called roll, which is upper case *Q*. Be careful not to get confused. Some terminals have two *?* symbols, one for APL and one for non-APL use. If you are unsure about which is correct, just experiment—you can't hurt a thing. Starting directly to the right of [1], type in

```
      [1] R←+/?6 & 6
```

and then hit (but *only* after checking line [1] very carefully to see if it is correct) EXECUTE. The computer will respond with [2] and you type ∇

```
[2] ∇
```

Then hit EXECUTE (or RETURN on some terminals) again.

Correcting Typing Mistakes

If you made a mistake on line [1], and you noticed it after pressing EXECUTE, then you can correct it very easily by typing in on line [2]

```
[2][1] R←+/?6  6
```

In short, you repeat the line number of the incorrect line and type in what it should be.

We have defined our function, but what is it? Type

```
      DICE
6
      DICE
7
      DICE
3
```

Generating Random Numbers

Generating Random Numbers

By now you have guessed that *DICE* is a function designed to give you the result of rolling a pair of dice. The basic element in the function is the primitive function roll, for which the symbol is *?* . Typing

```
      ?6
5
```

produces a *random number* between 1 and 6, typing *?*10, a random number between 1 and 10, and in general ?N produces a random number between 1 and N. The probability of any one specific number showing up is $1/N$. This is the monadic use of the aptly named roll function. Now we see why ? 6&6 produces two numbers from 1 to 6, each of which is equally likely to occur; this is the computer's simulation of rolling a pair of dice.

Our statement number [1] is now clear: *?* 6 6 gives us an array of two random numbers (statistically independent in fact), and +/*?* 6 6 gives us the sum of the two random numbers.

Function Headers

Function Headers

We are now ready to answer the question about the function header—the line that defines the function. (In this case the function header is ∇*R*←*DICE*.) In line [1], instead of writing *R*←+/*?* 6 6, we could have written: +/*?* 6 6. What's the difference? In either case, calling *DICE* produces the desired result in that the computer prints out the sum of a random tossing of two "computer dice."

The distinction between the two statements becomes apparent when we consider whether we want to use the result in some other APL statement. For example, we might wish to calculate *DICE*+2. Alternatively, we might want a function such that when the user types *HI*, the computer responds *WHAT IS YOUR NAME*. Now we do not want to *use* such a function in anything else; we just want the computer to type out *WHAT IS YOUR NAME*.

Interactive Programming

The *DICE* function is said to "return an explicit result," something we can use elsewhere, and the *HI* function does *not* produce an explicit result. If you want an explicit result, you need the header in this form: ∇ (Some *Variable* Name) ← (Name of Function). Somewhere in your function you will produce the explicit result that you want; that result must be stored under the same variable name as that used in the header. This is what we did in *DICE*. If instead of using *R* in the header we had written ∇*D*←*DICE*, then line [1] would have to read

```
[1] D←+/? 6 6
```

The Quote Function

To provide a contrasting example of a *no explicit result* function, consider the function *HI* mentioned above. Let's define *HI*:

Quote '

```
      ∇HI
[1] 'WHAT IS YOUR NAME?'          ← quote is upper shift K
[2] ∇
```

Now try typing

```
      HI                          ← you type
WHAT IS YOUR NAME?                ← computer responds
```

This example introduces another useful APL function, the quote ('), upper shift *K*. The characters between the quote signs are treated as just that: characters. When the computer comes across a line like [1] when in *execution* mode, it simply prints what it sees.

Monadic Function Headers

Monadic Function Headers

Let's return to functions with arguments. First, the header of a monadic (single argument) function is of the form

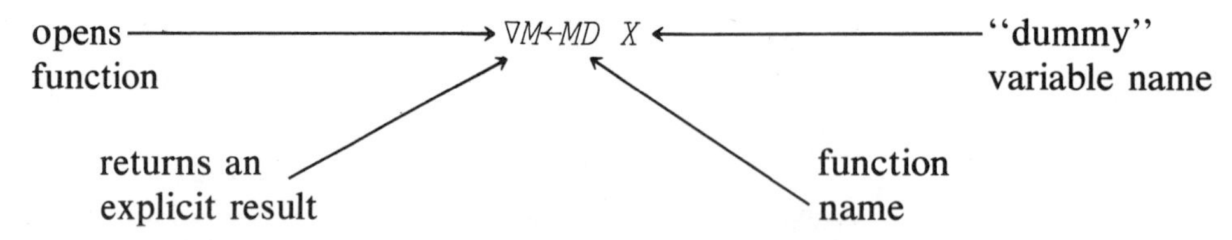

Why is *X* in the above header to the function *MD* called a "dummy" variable name? The reason is that in the header, *X* represents the argument that the function is supposed to operate on, and *X* appears later in the body of the function. But when you use the function *MD* on a variable *V*, or a variable *P*, or *A*, you would simply type in *MD V*, or *MD P*, or *MD A*.

Remember the median? We showed how to calculate it on page 40. Now we know how to make it a function. First, we need to decide on a name;

MD seems to be as good as any. Second, we need to decide whether we want an explicit result or not. Presumably, since the median is going to be used in some other calculations by someone at some time, it would be best to have it return an explicit result. Third, do we want the function to have arguments? Well, the answer to that is that if we have a vector, say *X*, in the computer we might want its median, and an easy way to say that is '*MD X*', so we want a monadic function. Recall that our first task is to enable the computer to decide whether *X* has an odd or an even number of elements in it (see page 40). But, somewhat more importantly, we want the computer to decide which way to calculate the median on the basis of whether *N* is even or odd. In short, we want the computer to do one computation if *N* is even and to branch to another computation if *N* is odd. As you might guess, this is a not so subtle hint that we are about to define a new APL operation; this one is called "branching."

Unconditional Branch

Suppose that in some APL routine you want the computer to go directly to line 6 if it gets to line 3. This is easily done by entering →6 on line 3.

```
[3]→6
```

Unconditional Branch →

The symbol → is the upper shift of the ← key. This is called "unconditional branch." In this case, the computer would know that if it ever gets to line 3 it must go directly to line 6 and execute it next.

Now for reasons that will become apparent in Chapter 7, it is a good idea to develop the habit of giving labels to lines to which the branch directs the computer to go. So if we label line 6 "*SIX*", for example, then the statement on line 3 could read [3]→*SIX*. In this case line 6 would look like this:

[6] *SIX*: {some APL instruction to be performed}

Line Labels

The computer reads from right to left, so when it encounters the symbol : in a line of a routine, it knows that everything to the left of : is the label for this line of instruction in the routine. We have simply named the line *SIX*. It now has a position number [6], and a label or name, *SIX*.

Conditional Branches

Let us now discover how to solve our problem. Recall the two ways in which we calculated the median (page 40). What we want to do is to go to one set of instructions or the other, depending on whether *N* is odd or even, and then to *exit* from the function. Let's see if we can program the following steps:

(a) determine whether *N* is even or odd;

(b) if *N* is odd, continue with the "odd" calculation and then exit from the function;

(c) if *N* is even, branch to the "even" calculation and then leave the function.

Consider (but do *not* type anything yet!):

```
        ∇ M←MD X
[1] N← ρ X
[2] →ODD×ι2|N
```

The last part of this line is clear enough: 2|N produces 0 if N is even and 1 if N is odd, since 2|N is the "two" residue of N, but what can we make of the first part? Let's look at the next section.

The Index Generator

Index Generator, ι

The symbol ι is the "index generator" (upper shift I). For any integer N, ιN produces the "index array" 1, 2, . . . , N. ι1 produces an array with only one element in it, namely 1. But what about ι0 ? This produces an "array" with nothing in it! The index generator, the monadic use of ι, unlike most of the other primitive functions in APL, *cannot* operate on an array; e.g., ιA , where A is an array, produces the result RANK ERROR.

RANK ERROR

Conditional Branches

If N is odd, 2|N produces 1, so line [1] produces an array of length 1 with a 1 in it, and multiplying this by ODD produces ODD. The first part of the statement says: Branch to the line labelled ODD. But what if N is even? Then ι2|N gives an array with nothing in it, and ODD multiplied by that produces an array with nothing in it still. APL interprets →(an array with nothing in it) as "ignore this silly statement and continue to the next line." So we have in line 2 a conditional branch statement; go to the line labelled ODD if N is odd, otherwise continue. So let *us* begin and put down the complete function. If you make a mistake typing any line, remember how to correct it: backspace to the beginning of the error and hit the linefeed or ATTN key, depending upon the type of terminal and computer system that you are using. On most CRT screens you just backspace and reenter the line.

```
        ∇ M←MD X
[1] N← ρ X
[2] XUP←X[⍋X]
[3] →ODD×ι2|N             ←if N is even, calculates M
[4] K← (ρ X)÷2              for even case, otherwise goes
                            to line ODD
[5] M←(+/XUP[K,K-1])÷2
[6] →0                    ←tells computer to leave function
```

```
[7]  ODD:M←XUP[1+(N-1)÷2]
[8]
```
← if N is odd, calculates M for odd case and then leaves function

Hold it!

Before typing the final ∇ and completing the function definition, check your entries very carefully. If a line has an error in it, then enter its line number (in brackets), type the correct version of the line, and hit RETURN. The computer will respond with the line numbers following the one you entered. If there are no more corrections, type in ∇ and RETURN; otherwise repeat the process.

Exit from Function

The first and third lines we have figured out. The three lines [2], [4], and [5] are copied from page 40, where we wrote down the routine for calculating the median M when N is even. But line 6 is a puzzle. Notice that the header of the function has no line number, and that line numbers start with 1. The function →0, or→(any line number not in function) tells the computer to "exit" from the function and go back to the point from which our function, *MD*, was called. In short, when the computer hits line 6 it knows it has finished calculating the median. It is best to stick to "→0" to exit from the function, since you may later want to add new lines to the function. The last two lines are modified from page 40, where we showed how to calculate the median where *N* is odd.

Let us try out our new median function. We put into the computer the *X* and *W* arrays we used earlier:

```
      X←1 2 ¯3 ¯4 5 ¯6 ¯7
      W←6 10 6 15 12 0 1 5
```

and now we can check the function out.

```
      MD X
¯3
      MD W
5.5
```

Correcting a Defined Function

Correcting a Defined Function

What if you enter *MD X* and you do *not* get ¯3? You've made a mistake! Let's see how to correct it. First, relist the variables *X* and *W* and make sure that they are correct. If they are, continue; if not, redefine *X* or *W* and retry the function.

But suppose you are sure the mistake is in the function *MD*? In order to check it, you will need to display it. To do this, we type in:

Display, ⎕

```
        ∇ MD[⎕] ∇
        ∇ M←MD X
[1] N←ρ X
[2] XUP←X[⍋X]
[3] →ODD×ι2|N
[4] K←(ρ X)÷2
[5] M←(+/XUP[K,K+1])÷2
[6] →0
[7] ODD:M←XUP[1+(N-1)÷2]
        ∇
```

← this line "opens" the function, displays it (that's what [⎕] does), and closes it. ⎕ is typed in by upper case *L*.

← this is the computer's response to your command to display the function

Suppose you discover that you made some typing errors in the function. In Chapter 5 we will show you in detail how to correct errors, but for now we will give you one drastic, but surefire, way to do it. If we remove the function altogether, we can completely redo it—this time very carefully. To remove the function, type

)*ERASE* *Command*

```
        )ERASE MD
SI DAMAGE
```

← this statement is the computer's response. It warns you that you have erased a suspended function. SI stands for State Indicator.

Suspended Functions

You might also get a statement indicating there is a syntax, value, or some other error in a line of your function when you try to execute the function with an error in it. For example,

```
        MD X
VALUE ERROR
MD[1]N←RX
       ∧
```

← this is the computer's response telling you the first line and first position of your first mistake

Removing a Suspended Function

The result of this error is to prevent further execution of the function. It is said to be suspended. At this time, it is best to remove the suspension. This is easy—type

```
        →
```

← this is upper case "←"

In order to check whether you have cleared the suspension, type

)*SI* *Command*

```
        )SI
```

If you get no apparent response (actually a blank line) all is well. But if you get something like

```
MD[3]*
```

type in the → instruction again. Keep doing this until you get a blank response to the system command)SI.

Now erase the incorrect function and start again. Remember, you cannot correct the function simply by redefining it unless you first erase the old function. (This is not strictly correct—you could do it if you gave the correct function a new name (i.e., one that differs from the old name).)

One last word: if you think that the MD function is a lot of fuss about a simple operation, you're right. Later on we will show you how to calculate the median in one short line. But in writing out the routine MD you learned a lot about APL that will be very useful in the following chapters, and that, of course, is what this book is all about.

Summary

Sample Median, *M* (of an array *X*): that value such that half the observations in *X* are less than or equal to *M* and half are greater or equal to it.

Residue function, |: in its dyadic form it is written *A*|*B*. For positive values of *A* and *B*, the *A* residue of *B* is the remainder of dividing *B* by *A*.
The monadic use of | yields the absolute value.

Indexing, [I] (of an array *X*): elements of an array *X* can be selected by specifying their position in the array; the positions are labelled from 1 to ρ*X* starting from the left. X[I] gives the value of the number in the array X in the *I*th position from the left.

Grade up, ⍋ (upper shift M, backspace, upper shift H) (of an array X): gives an array, the first element of which is the index position of the smallest number in X, the second element is the index of the next largest, and so on.

Grade down, ⍒ (upper shift M, backspace, upper shift G) (of an array X): gives an array, the first element of which is the index position of the largest number in X, the second element is the index position of the next smallest, and so on.

Catenate, "," (the comma): forms one array out of its arguments; A,B is an array with the elements of A listed first where A or B can be any one-dimensional array or scalar.

Roll, ? (upper case Q): ?N, where N is an integer that generates a random integer from 1 to N with probability 1/*N*.

Index Generator, ⍳ (upper shift I): monadic function, ⍳N, N an integer, produces the "index" array 1, 2, 3, . . . , *N*.

Quote, ' (upper shift *K*): all keyed entries between a pair of quotes are treated as characters and not executed as APL functions.

Del, ∇ (upper shift *G*): the instruction that is used when entering or exiting from function definition mode.

Niladic function: a function which has no arguments; it carries out a specified procedure when called.

Function Headers: first line of a function that gives the function name; specifies whether function produces an explicit result or not, and contains either zero, one, or two arguments. Thus:

∇ *HI*	niladic (no argument) function, no explicit result produced
∇ *A GO B*	dyadic function, no explicit result produced
∇ *R←FNC*	niladic function, explicit result produced
∇ *R←FNC X*	monadic function, explicit result produced
	(All six combinations are possible)

In function definition mode, a line in a function can be corrected by typing in the line number and the correct version of the whole line. For example, in correcting line [1] at line [3], you would type: [3] [1](correct APL statement).

Unconditional Branch, →: used within a function to alter execution path to a different line. Thus:

```
[3] →6
```

or

```
[3] →LABEL
[7] LABEL: (an APL expression)
```

Line Label: a name for a line in a function. When the second of the above versions of line 3 is executed, the next line to be executed will be number 7 with the label *LABEL* rather than number 4.

Conditional Branch: branching to a function line on the basis of the "condition" of some APL expression. Example:

```
[3] →TRUE×⍳2|N
[4] 'THE CONDITION IS FALSE'
  .
  .
  .
[6] TRUE: (another APL expression)
```

If 2|*N* takes the value 1, the next executable statement is line [6] labelled *TRUE*. If 2|*N* takes the value 0, next executable statement is [4].

Quad, ⎕ (upper shift *L*): when used as illustrated below with the function *MD* it displays the function *MD*:

∇ *MD*[⎕] ∇

)*CLEAR* : a system command used to remove all existing functions and variables from the existing workspace. The workspace can be thought of as the part of the computer which is assigned to the user for his computation procedures (both stored data and functions).

)*ERASE* : (*FUNCTION NAME*): a system command that enables you to erase a function from your workspace.

)*SI* : system command called state indicator. If you have functions which cannot complete their operation, they are said to be suspended. The command)*SI* will indicate which functions are suspended and at what line in the function the suspension occurs.

To clear a suspended function, type → as often as needed to get a blank response from the command)*SI*.

Exercises

APL Practice

1. Let's examine in some detail the use of the new APL functions presented in this chapter. For *X*←¯5+⍳11, compare the results of the operations in each question.

(a) 1|*X*

(b) 0|*X*

(c) 2|*X*

(d) ¯2|*X*

(e) 10|*X*

(f) *X*|0

(g) *X*|¯10

(h) *X*[⍳3]

(i) *X*[2,3]

(j) *X*[⍳2]←0&0 and then ask for *X*

(k) *X*[1 1 2 3]

(l) *X*[⍳4]

(m) *X*[⍋⍋*X*] and compare to *X*[⍳ ⍴*X*]

(n) *X*[0]

(o) *X*[12]

(p) *P*←⍳4 and then *X*[*P*]

(q) *X*|*X*

2. Since you know that the meaning of ?6 6 and 6 6 is equivalent to 2⍴6, why not combine them into ?2⍴6? Try:

(a) ?3 ⍴ 6

(b) ?3 ⍴ 0

(c) ?3 ⍴ 1

(d) ?1 ⍴ 1

(e) What is the difference between `3ρ?6` and (a)?

(f) What is the difference between `ρX,ρY,(ρX),(ρY)` and `ρX,Y` for any arrays *X* and *Y*?

3. Indexing can be used with the variables defined as a list of characters. Consider the following code game. Let

```
AL←'I WILL MEET YOU TOMORROW AT SEVEN'
```

(a) `AL[8 26 22 26]` tells whom you will meet

(b) `AL[22,9,29,17,26,15,22,26,33,17]` tells where

(c) `AL[10,26, 27]` says what you will do

(d) `AL[29,30,31,32,33]←'SIX'` changes the time of the meeting

4. Let `Z←¯5+⍳11`

(a) `((6ρ1),(5ρ0))/Z`: picks the six first elements of *Z*, and is equivalent to `Z[⍳6]`.

(b) Show how to pick the last four elements of *Z*.

(c) For any array *Z* write an APL expression that will pick the middle n elements. What if `ρZ` is even?

(d) Is `Z[⍒⍒Z]` equal to `Z[⍋⍋Z]`?

(e) Suppose you want to delete the seventh element of *Z*. How would you ask the computer?

5. Can you discover, with the computer's help, which of the following series converge to a limit and which diverge? If the series converges, can you determine what the limit is? How accurate is your answer? Compare your result with a table of limits of series.

(a) $1 - \frac{1}{2} + \frac{1}{3} - \frac{1}{4} \cdots$

(b) $\frac{1}{2} + \frac{2}{3} + \frac{3}{4} + \frac{4}{5} + \frac{5}{6} + \cdots$

(c) $\frac{5^0}{1} + \frac{5^1}{1} + \frac{5^2}{2} + \frac{5^3}{3} + \frac{5^4}{4} + \cdots$

(d) $\frac{1}{2^0} + \frac{1}{2^2} + \frac{1}{2^3} + \frac{1}{2^4} + \cdots$

(e) $\frac{1}{2.3} + \frac{2}{3.4} + \frac{3}{4.5} + \frac{4}{5.6} + \cdots$

(f) $e^{-1} + e^{-2} + e^{-3} + e^{-4} + \cdots$

(g) $\left(1 + \frac{1}{1}\right)^1 + \left(1 + \frac{1}{2}\right)^2 + \left(1 + \frac{1}{3}\right)^3 + \cdots$

(h) $\frac{4}{2.3} + \frac{5}{3.4} + \frac{6}{4.5} + \frac{7}{5.6} + \cdots$

(i) $(-1)^1 \left(\frac{1}{1+1}\right)^1 + (-1)^2 \left(\frac{1}{2+1}\right)^2 + (-1)^3 \left(\frac{1}{3+1}\right)^3 + \cdots$

(j) $1 \cdot \ln\left(1 + \frac{1}{1}\right) + 2 \cdot \ln\left(1 + \frac{1}{2}\right) + 3 \cdot \ln\left(1 + \frac{1}{3}\right) + \cdots$

(k) $\frac{e^1}{1+1} + \frac{e^2}{2+1} + \frac{e^3}{3+1} + \frac{e^4}{4+1} + \cdots$

(l) $\frac{\log 1}{1+\log 1} + \frac{\log 2}{2+\log 2} + \frac{\log 3}{3+\log 3} + \cdots$

6. Let the list $M \leftarrow 3\ 3.2\ 3.5\ 3.6\ 3.6\ 3.7\ 4\ 4.2\ 4.5\ 5.1\ 5.3\ 5.5$ represent the annual U.S. money supply in hundred billions of dollars for a 12 year period. Define a new list consisting of the first differences of the elements of M. That is, for $i = 2, 3, \ldots, 12$, let $DM_i = M_i - M_{i-1} = 3.2 - 3, 3.5 - 3.2, 3.6 - 3.5, \ldots$, eleven terms in total. What are the mean and variance? How about the list $(M_i - M_{i-1})/M_{i-1}$, $i = 2, 3, \ldots, 12$? The list $\log M_{i-1}/\log M_i$, $i = 2, \ldots, 12$?

(*Hint:* One suggestion for finding the first differences of a list is:

```
     ∇ DM←DEF X;M1;M2
[1] M1← (0,((ρX)-1)ρ1)/X
[2] M2← ((((ρX)-1)ρ1),0)/X
[3] DM←M2-M1
[4] ∇
```

Notice that `DM1←DEF M` produces the first differences of `M`. `DM2←DEF DM1` would produce the second differences of `M`, `DM3←DEF DM2` would produce the third differences of `M`, and so on. What are the mean values of the differences `DM1`, `DM2`, `DM3` . . .?)

7. Here are some important inequalities and identities.

(a) The Cauchy-Schwarz inequality

$$(\Sigma_{i=1}^N x_i y i) \leq (\Sigma_{i=1}^N x_i^2)(\Sigma_{i=1}^N y_i^2)$$

(b) Hölder's inequality

$$(\Sigma_{i=1}^N x_i y_i) \leq (\Sigma_{i=1}^N x_i^p)^{1/p}(\Sigma_{i=1}^N y_i^q)^{1/q}$$

for any p and q such that $1/p + 1/q = 1$, $p > 1$.

(c) Minkowski's inequality

$$[\Sigma_{i=1}^N (x_i + y_i)^K]^{1/K} \leq (\Sigma_{i=1}^N x_i^K)^{1/K} + (\Sigma_{i=1}^N y_i^K)^{1/K}$$

for $x_i y_i \geq 0$ and $K > 1$

(d) For $\Sigma_{i=1}^N a_i$ and $\Sigma_{i=1}^N b_i$ convergent series of positive numbers such that $\Sigma a_i \geq \Sigma b_i$, we have

$$\Sigma_{i=1}^N a_i \log(b_i/a_i) \leq 0$$

(e) The Lagrange identity

$$(\Sigma_{i=1}^N x_i^2)(\Sigma_{i=1}^N y_i^2) - (\Sigma_{i=1}^N x_i y_i)^2 = \Sigma_{i<j}^N (x_i y_j - x_j y_i)^2$$

In some advanced statistics courses, these inequalities and identities are proven by algebraic methods. However, here is a way to use the computer to be fairly certain that the result can be proven algebraically.

Pick four integers at random (*Hint:* use ?); say you get 4 9 10 15. For each integer pick a random sample of numbers for list x and another for a list y, where the number of elements in each list is in turn 4 9 10 15. Check to see that the suggested inequalities hold with each sample. Repeat the same experiment using different random lists x and y each time.

If you discover no reversal of the inequalities (or violation of the identity), you may reasonably suspect that the result can be proven algebraically, and hence that it is a mathematically correct statement.

The above procedure is known as a "sampling experiment," and provides the basis for what are known as "Monte Carlo" procedures. Monte Carlo procedures often are used to solve mathematical problems that are difficult to solve by analysis.

8. Enter the function *MD X* (page 48) into your workspace.
   (a) Display the function.
   (b) Describe what the function instructs the computer to do at line [3].
   (c) Add the line 'The Median of X is' between lines [5] and [6].
   (d) Execute the function and press the ATTN key before you get any answer (i.e., while the computer is in execution mode). Press the command)*SI* .
   (e) Get out of the suspension.
   (f) What would the computer do if you changed line number [6] of the original function into →1 and the number of elements in X is even?

Statistical Exercises

1. Let C←¯1 + ?100 ρ 2 represent the outcome from tossing a coin 100 times. The random variable C takes on the value 0 if a "head" and 1 if a "tail." If a head occurs you win one dollar, while if a tail occurs you lose one dollar. Calculate how many dollars you expect to win or lose if you play this game with the computer. Try it four times. What would you pay to play this game?

2. The standard deviation of the binomial distribution with parameters n and p is given by the formula $S = \sqrt{np(1 - p)}$, where n is the sample size and p is the probability of success in a single trial. For $n \leftarrow 100$ and $p \leftarrow (\iota 10) \div 10$ find the value of p for which S is maximum. Is there a minimum?

3. Your friend from Texas called you up and asked you to find out what the average price is of used cars sold in N.Y. You collect a small sample of New York used car prices as shown in the following table.

P	2.5	3	3.5	4	4.5	5	5.5	6	6.5	7	7.5	8	8.5	9
F	0	2	5	4	6	15	14	18	11	8	9	4	3	3

P is the price of the car in thousands of dollars rounded to the nearest 0.5 thousand, and F is the frequency, i.e., how many cars you observed at each respective price. You have a sample size of 102. Calculate the sample mean, variance, and standard deviation. Plot the histogram of P vs. F. How would you convey to your friend, who does not know any statistical theory, the information you have collected? How would you advise her?

4. One hundred cans of floor wax, randomly selected from a large production lot, have the following net weight in ounces with the corresponding frequencies:

weight	19	19.2	19.4	19.6	19.8	20	20.2	20.4	20.6	20.8	21
frequency	0	3	8	9	16	20	15	12	4	5	6

(a) Find the sample mean weight ($\bar{x}$).

(b) Find the sample median (Mx) and compare its value to the mean. Explain the difference.

(c) What is the value of the sample standard deviation S?

(d) What percentage of the sample falls into each of the following intervals:

1. $\bar{x} \pm 1S$
2. $\bar{x} \pm 2S$
3. $\bar{x} \pm 3S$

(e) The variation in production by net weight is "acceptable" when the mean value is 20 ounces and 95% of the cans produced contain no more than 20.3 nor less than 19.7 ounces. Can we conclude that the production run is acceptable on the basis of this sample?

5. The formula $P(y_i) = (0.1)(y + 1)$ gives the probability of $y = 0, 1, 2, 3$ occurring.

(a) Graph $P(y_i)$.

(b) Verify that $P(y)$ is a probability distribution function, that is, that $\Sigma_{i=1}^{4} P(y_i) = 1$.

(c) Calculate the cumulative probability distribution function of y and graph it.

(d) Find the expected value of y.

(e) What is the value of $P(y < 3)$. That is, what is the probability of y being strictly smaller than 3?

(f) What is the probability of y being smaller than 0?

(g) What is the probability of y being smaller than 10?

(h) Calculate the standard deviation of y.

(i) Calculate the third and fourth moments about the mean.

6. Suppose u is a random variable that assumes the integer values from one to ten, each with equal probability of 0.1. Use the computer to do the following:
   (a) Pick 20 random samples, each of size 30, from the distribution of u.
   (b) Calculate the mean value and sample standard deviation of u for each of the 20 samples.
   (c) Print the vector of mean values and call it U.
   (d) Calculate and print the sample standard deviation of the 20 mean values. (Call it SU).
   (e) Calculate and print the percentage of the samples whose means fall into the intervals

      1. $5.5 \pm 1Su$
      2. $5.5 \pm 2Su$
      3. $5.5 \pm 3Su$

      where 5.5 is the mean of the distribution u.
   (f) Calculate and print the percentage of the samples for which the intervals

      1. $\bar{u}_i \pm 1Su_i$
      2. $\bar{u}_i \pm 2Su_i$
      3. $\bar{u}_i \pm 3S\ddot{u}_i \qquad i = 1, 2, \ldots, 20$

      include the number 5.5.

   What is the connection between the results in (e) and (f)?

5

Some More Statistics

The major objectives of this chapter are to enable you to calculate a few more simple statistics and probabilities, practice your computer skills, and learn some more useful APL functions.

Ravel

Ravel, ,

First, let us clear up a small difficulty with the routine we used to calculate the arithmetic mean (see page 20). We calculated *N*, the number of observations, by *N*←ρ*X* where *X* is the array of numbers whose mean we wanted. What if *X* is a scalar, not an array? In this special case, we would like the mean of *X* to be *X*; but we will not get that result by using *N*←ρ*X*, since ρ*X* is "blank." There is an easy way out: our friend the comma (,), used this time as a monadic function, is called *ravel*. Ravel simply makes its argument an array, whatever its shape was to start with. Thus if *X* is scalar, then ,*X* is an array of length one. Try the following:

```
X←3

(+/X)÷ρX
```

Nothing seems to have happened. There is no response since division by "blank" is undefined. Now try

```
(+/X)÷ρ,X
```

```
3
```

Now we have the correct result.

5.1 Some Basic Statistics

APL Routine to Compute Mean, Median, Variance, Standard Deviation

Let's consider a routine that will calculate a number of basic statistics, say the mean, median, variance, standard deviation, the range, and so on. To increase our understanding of how to define and use functions, let us define a function which has a single argument which is an array *X*, does the required calculations, and prints out the results, but does not yield an explicit result.* That is, the routine performs a series of operations but does not return a value that can be used in other calculations. Type in

```
      ∇DSTAT X
[1] R← (MAX←X[ ρ X])-MIN← (X←X[⍋X])[1]
[2] SD← (VAR← (+/ (X-MEAN← (+/X)÷N)*2)÷ (N← ρ X)-1)*0.5
[3] MD←( +/|X-MEAN)÷N
[4] MED←0.5×+/X[(⌈N÷2),1+⌊N÷2]
[5] 'SAMPLE SIZE'
[6] N
[7] 'MAXIMUM'
[8] MAX
[9] 'MINIMUM'
[10] MIN
[11] 'RANGE'
[12] R
[13] 'MEAN'
[14] MEAN
[15] 'VARIANCE'
[16] VAR
[17] 'STANDARD DEVIATION'
[18] SD
[19] 'MEAN DEVIATION'
[20] MD
[21] 'MEDIAN'
[22] MED
[23]
```

* This routine is a modified version of one appearing in Smillie (1969, p. 16).

Hold it. Don't type the final ∇ until you are sure that you have typed everything just as it's written above! If you made a mistake in any line, correct it as you did in Chapter 4. When you are sure that the function is correct, close it out.

```
[23] ∇
```

Let's see if the function works. Type in our old friends the X and W arrays.

```
      X←1 2 ¯3 ¯4 5 ¯6 ¯7
      W←9 10 3 8 12 0 1 5
      DSTAT X
SAMPLE SIZE
7
MAXIMUM
5
MINIMUM
¯7
RANGE
12
MEAN
¯1.7143
VARIANCE
19.905
STANDARD DEVIATION
4.4615
MEAN DEVIATION
3.7551
MEDIAN
¯3
      X
1 2 ¯3 ¯4 5 ¯6 ¯7
      DSTAT W
SAMPLE SIZE
8
```

```
MAXIMUM
12
MINIMUM
0
RANGE
12
MEAN
6
VARIANCE
19.429
STANDARD DEVIATION
4.4078
MEAN DEVIATION
3.75
MEDIAN
6.5
      W
9 10 3 8 12 0 1 5
```

Correcting a Function Line

It seems as if our function works, but what if yours did not?

If your function did not produce the same results as ours, but otherwise appeared to work, then first of all check that the variables *X* and *W* are exactly the same as ours. If they are, but you still got some strange results or statements about errors in the routine, then you know that you have made a mistake in defining the function.

You know that you can always use the system command

)ERASE

```
      )ERASE DSTAT
```

to remove the function. But there is a better way!

First, in case your function is suspended, remove the suspension in the way that we showed you in the last chapter. Then check with the)*SI* system command to ensure that the suspension is completely removed.

Correcting a Function Line

The next thing to do is to see how we can correct a line in our function. Let's suppose that you have discovered your mistake lies in line [3]. You want to correct just that line. Here's how to do it. Type in

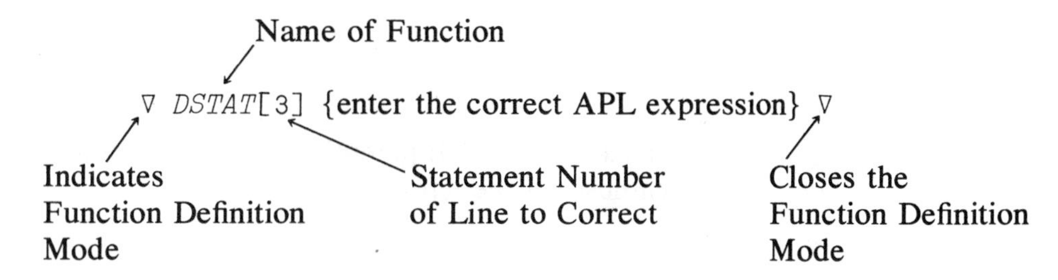

and then hit the RETURN or EXECUTE key. For example:

```
∇ DSTAT[3] MD←(+/|X-MEAN)÷N∇
```

If you have several lines to correct you can use the above procedure repeatedly.

Now that you have a working function, let's see how it works. Consider the first line:

```
[1] R←(MAX←X[ρX])-MIN←(X←X[⍋X])[1]
```

Starting from the right, the first group of symbols needing our attention is

```
(X←X[⍋X])[1]
```

Let's skip the [1] for a moment. First, we sort the array *X*, and then we replace the old *X* array by the new sorted array *X*. Thus `(X←X[⍋X])` results in an array. `X[⍋X]` reorders the elements of `X` in ascending order, using the grade up function `⍋`, and the reordered elements are replaced in the `X` array. In short, `X` is now an array containing the original elements of `X` rearranged in ascending order. Now the `[1]` in

```
(X←X[⍋X])[1]
```

gives us the first and, therefore, minimum element of `X`. This result is promptly placed into a variable called `MIN` by

```
MIN←(X←X[⍋X])[1]
```

The next function, −, instructs us to subtract *MIN* from the next argument, `(MAX←X[ρX])`, and to put the result in *R*. Note that (proceeding from right to left) after `MIN` has been calculated, `X` is in ascending order. The maximum of `X`, which is its last element, is therefore given very simply by `X[ρX]`. We now have the "range" of the array, i.e., the difference between the maximum and minimum of `X`.

After line [1], line [2] should be easy. Reading from the right, after `)*0.5` we have:

`(N←ρX)-1`	stores number of observations in N and computes $(N - 1)$
`(+/(X-MEAN←(+/X)÷N)*2)`	computes $\Sigma_i(x_i - \bar{x})^2$
`(X-MEAN←(+/X)÷N)`	computes the array X-MEAN

Finally, raising the whole expression to the 0.5 power (computing the square root) yields the standard deviation. In tracing this through, remember to match up the parentheses in *pairs*. For example, in the first term above, one does not have $(N - 1)^{1/2}$ as you might at first think, because

```
(N←ρX)-1)*0.5
```

involves right and left parentheses that are not matched until you reach the parenthesis to the left of *VAR*.

Absolute Value, Floor, and Ceiling

Absolute Value Function, |

Line [3] involves the *monadic* use of |, which produces the absolute value of a number (discussed in the previous chapter). Try keying in

```
      |3.2
3.2
      |¯3.2
3.2
      |-2.6
2.6
      XABS←|X
      XABS
1 2 3 4 5 6 7
```

Floor, ⌊ *Ceiling,* ⌈

Line [4] shows the quick way to get the median and introduces two new, but related, functions: Floor ⌊ and ceiling ⌈, both used here as monadic functions. Floor ⌊ rounds its argument down to the next lowest integer, whereas ceiling ⌈ rounds up to the next highest integer. Floor ⌊ is keyed in as upper shift *D*; ceiling as upper shift *S*. For example,

```
      ⌊ 2.3 2.5 2.7
2 2 2
      ⌈ 2.3 2.5 2.7
3 3 3
```

If *N* is even, say 6, then the index function

```
      X[ ( ⌈N÷2),1+ ⌊ N÷2]
```

yields the third and fourth elements of *X*, which was reordered into ascending order in line [1]. If *N* is odd, say 7, then

```
      X[(⌈N÷2),1+⌊N÷2]
```

produces the fourth (middle) element of *X* twice. Plus reduction followed by multiplication by 0.5 gives the correct answer in both cases.

Line [4] is clearly a much better way of calculating the median than is the procedure we used earlier. Since that approach has now served its purpose, you should use line [4] or something like it for calculating a median. But the experience provided an important lesson: the *first* way you think of doing something is probably not the most efficient way—so you might think of some alternatives before proceeding with your attempt.

5.2 Dummy, Local, and Global Variables

You may have wondered why we printed out the arrays *X* and *W* after using the function *DSTAT* in §5.1. However, if you think about what was done in line [1] in *DSTAT*, you might instead wonder why it is that although we rearranged the arrays *X* and *W* into ascending order in line [1], they were in their original order when we printed them out. The explanation for this result is that the variable *X*, which occurs in both the header and the body of the function, is "local" to the function. So anything that happens to an *X* *inside* the function does not affect any *X* that exists outside the function. This idea seems intriguing, so let's pursue it.

Up until now, you have become accustomed to the idea that if a variable is redefined in some function, then the old definition is lost and only the new one prevails. For example,

```
      N←6
      N
6
      N←(N+1)*2
      N
49
```

Global Variables

and so on. As we shall see, the variables we have been using so far are known as "global variables" or, rather, *the way in which we have defined and used* these variables makes them global variables. The values represented by the variable names which are globally defined can be called and used by any APL expression at any time.

Dummy Variables

Dummy Variables

The variables that are used in a function header as the *arguments* of the function are known as dummy variables; these variables are in fact defined *only* within the function itself and nowhere else! For example, if you start with a clear workspace, we can illustrate this idea with an experiment. Type

```
      )ERASE A B
      ∇C←A DUM B
```

This is to ensure that *A*, *B* are not defined as variables anywhere in your workspace.

```
[1] C←A←1+B←3 ∇
      A
VADUE ERROR
      A
      ^
      B
VALUE ERROR
      B
      ^
      A DUM B
VALUE ERROR
      A DUM B
         ^
```

This sets up a function with A, B as arguments in the function.

Value Error

This indicates that after defining the function *DUM*, both *A* and *B* are still undefined in the computer workspace.

We get a *VALUE ERROR* because *A*, *B* are undefined outside the function *DUM*, so that we cannot use undefined variables as arguments for the function.

Suppose we assign zero to both *A* and *B*, thereby defining

```
      A←B←0
      A DUM B
4
      A
0
      B
0
```

Here we use the function and it produces the value 4. In this case the result is completely independent of whatever we have defined *A* or *B* to be outside the function.
So, even after execution, *A*, *B* still have the values assigned to them initially. Although the function produced the expected result of 4, *A* and *B* have *not* been changed to 4 and 3, respectively.

We see from this experiment that the variables used as *arguments* in the *definition* of a function are distinct from global variables. When *A* and *B* were set to 0 by use of the replacement statement, they were defined globally. The variables in the *definition of a function* serve a role distinct from that of variables generally.

Consider the mathematical statement defining the summation of a set of n numbers $(a_1, a_2, \ldots, a_n)$:

$$\Sigma_1^N a_i$$

In this definition, the index i is a dummy variable also—we could use any letter and get the same result. Thus

$$\Sigma_1^N a_i = \Sigma_1^N a_j = \Sigma_1^N a_r$$

and so on. Correspondingly, in our function *DUM*, the variables *A* and *B* used in the *definition* served the role of dummy variables. They merely indicate each role that the first and second arguments play in the function. If we substituted any other two variable names in the function *definition* we would get the same results on execution. But when we come to the execution of a function, globally defined variables must be used in calling the function; otherwise a value error will be generated. For a function to perform its assigned task it must have some values on which to operate. Assigning values to variable names by the assignment operator ← does just that.

The above example of summation indices gives us an idea why we might want to distinguish global and dummy variables. First of all, we see that we will want our function definitions to be general in the sense that we would like to define the function *once* and then be able to use it with any variables we choose. When we define the function we will have to use some symbols to represent the arguments. Consequently, you can see the benefits of deciding that variable names specified in function definitions should have no meaning outside the definition of the function, and that to *use* the function global variables must be specified.

When you become more expert at computing you will discover that you will have to handle a large number of variables, and that it is very easy to forget what is what. A part of the problem is that it is very easy to give the same variable name to two different variables—and you already have discovered that the second assignment replaces the first. In any case, you can appreciate how confusing it all can be. Consequently, it is a relief to know that when you define a function and the use of its arguments, you will not affect the values of any global variables that you have already defined and want to keep.

What about variables that appear in the body of a function, but *not* in the header? These are also global variables and they will be defined when the function is *executed,* but not before. For only when the function is executed are the operations defined in the function actually carried out. Before then the operations have been *defined,* but not yet used.

Local Variables

Local Variables

The idea of variables that appear as arguments in the header of the function, so that they are defined only in the context of the definition of the function, can be extended to "local variables." Local variables are defined in the header by separating the specification of the local variables from the function header (and from each other) by semicolons. For example, consider the following headers:

```
∇R←A GET B;X;Y

∇A BY B;X;Y

∇ FUNC;X
```

X is a local variable in each of the above functions and *Y* is a local variable in the first two. Note that local variables can be defined with an explicit functional result (function *GET*), without an explicit functional result (function *BY*), and even for functions without arguments, as in *FUNC*.

Local variables differ in several respects from dummy argument variables. When a function is *used,* that is, when the computer is instructed to perform the mathematical operations specified by the function definition, then *global* variables must be inserted where the dummy variables appear—a one-argument function needs one global variable to operate on; a two-argument function needs two global variables to work on. But the local variables, which are specified in the header, are *defined* in the *body* of the function and do not have to (indeed, *must not*) be supplied by you when using the function. For example, one would call the above functions by

```
      P GET Q,   or
      S BY T,   or
      FUNC,
```

where *P*, *Q*, *S*, *T* are global variables defined elsewhere in the computer before these functions are executed.

Consider a use of a local variable:

```
      X←1 2 ¯3 ¯4 5 ¯6 ¯7
      ∇ FUNC;X
[1] X←2
[2] X ∇
      FUNC
2
      X
1 2 ¯3 ¯4 5 ¯6 ¯7
```

We see from this example that the globally defined *X* and the *X* defined locally to the function *FUNC* are two entirely different variables. The specification of local variables lets you extend the advantages of the dummy variables to more variables.

Dummy variables that appear as arguments in the header of a function when *defining* it enable one to define the mathematical operations to be performed. *Use* of the function requires substituting previously defined global variables.

Local variables enable us to reuse variable names within a function without affecting whatever definitions these variable names may have elsewhere.

Summary

Ravel, ",": a monadic function that makes its argument a one-dimensional array.

DSTAT: a routine in this book defined to calculate maximum and minimum values, range, mean, variance, standard deviation, mean deviation, and median.

A statement, say the third, inside a function routine can be corrected by:

∇ *DSTAT*[3] {correct APL expression} ∇

Absolute value, | (upper shift *M*): a monadic function that gives the absolute value of a number.

Floor, ⌊ (upper shift *D*): rounds number down to the next lowest integer.

Ceiling, ⌈ (upper shift *S*): rounds a number up to the next highest integer.

Dummy Variable: variable names used in the definition of a function's arguments, are not defined outside the function.

Global Variables: variables defined by the assignment operator, can be used by any APL expression or routine.

Local Variables: variables defined only within a function, are designated in the header.

EXAMPLE:

```
    A←3

    ∇ R←X FNCT Y;Z1;Z2;Z3
[1] B← some APL expression
[2] R← some APL expression
[3] ...
   ... ∇
```

Result Variable:	*R*
Dummy Variables:	*X*, *Y*
Local Variables:	Z1, Z2, Z3
Global Variables:	*A*, *B*

Exercises

APL Practice

1. Let `X←(?10 ρ 10)÷2`
Drill
(a) `4 ρX`
(b) `( ρX) ρX` which is of course *X*
(c) `30 ρ 'X'` and try `3+30 ρ 'X'`. (You get a Domain Error because *X* is an array of characters, not numbers.)
(d) `ρ(ρX)ρX`
(e) Is `⌈3.4` the same as `⌊ 3.4+1`?

(f) `⌊ X`

(g) Is `⌈X` the same as `⌊ X+1`?

(h) Is `⌊⌊X` the same as `⌊X`?

(i) Is `⌈⌈X` the same as `⌈X`?

(j) What do you get by using `⌊/X`; or `⌈/X`? Compare `⌊\X` and `⌈\X`.

(k) Is `X-1|X` the same as `⌊X`?

(l) Is `X+(1-⌊|X)` the same as `⌈X`? What if some X_i are integers?

(m) Is `(⌈/X)` the same as `(X[⍋X])[⍴X]`?

(n) Find the two largest values of X.

(o) Find the third smallest value of X.

2. Let `W←(¯5+⍳9)÷2`, `YY←1|W`, and `Z←|W`, i.e., Z is the array of the absolute values of W.

(a) Compare `⌈W` to `⌈Z` and `⌈YY`.

(b) Compare `⌊W` to `⌊Z` and `⌊YY`.

(c) Compare `Z` to `(W*2)*.5`.

(d) Compare `-Z` to `-(W*10)*.1`.

3. Consider the polynomial $f(X) = 10X - X^2$, where `X←¯51+⍳101`, i.e., X takes only the integer values from ¯50 to 50. For these values of $X, f(X)$ might be either positive or negative.

(a) Find the maximum value of $f(X)$.

(b) Find the values of X for which $f(X)$ is positive.

(c) Display all the negative values of $f(X)$.

4. Using `X←(¯51+⍳101)÷10` and $f(X) = -X^4 + 3X^2 + 1$, find the two local maxima of $f(X)$ as well as the one local minimum.

5. For any array V verify that the following APL expressions are equivalent.

(a) `(+/V)÷ ⍴V`

(b) `+/V÷ ⍴V`

(c) `V+.÷ ⍴V`

(d) `1+.×V÷V+.*0`

6. Put the following function into your workspace:

```
      ∇ J←JOHN I;A;B;C;D
[1] A←1×I
[2] B←2×I
[3] C←3×I
[4] D←4×I
[5] J←A+B+C+D
[6] ∇
```

(a) What does the function do?
(b) Which variables are local?
(c) Which variables are global?
(d) Which variables are dummy?
(e) Which variable contains the result?
(f) Is the function monadic, niladic or dyadic?
(g) Does the function give you an explicit result?

7. Consider the polynomial $f(X) = -1 \cdot X^0 + 2X^1 + 3X^2 + 5X^3 - 10X^4 + 2X^5$. Let `B←¯1 2 3 5 ¯10 2` be the array of coefficients and `P←⍳5` the array of the exponents of X. Then `B+.×K*P` evaluates the polynomial for $X = K$.
   (a) Evaluate the polynomial for $K = -5$ and $K = 3$.
   (b) Does this polynomial have a root for $-10 < X < 10$?

8. Refer to exercise 7. Suppose you discover that there is one root between 0 and 1. Write a program that will find the root to the nearest thousandth.

Statistical Applications

1. Use your *DSTAT X* function and the samples

```
W←1 2 3 4 5 6 7  8  9  10 11

Z←1 2 4 5 5 6 7  7  8  10 11

Y←1 1 1 2 2 6 10 10 11 11 11
```

   to verify that
   (a) All three samples have the same number of elements, the same maximum, the same minimum, the same range, the same sample mean, and the same median.
   (b) The median is equal to the mean for all three samples.
   (c) Z has the smallest mean deviation, standard deviation, and variance.

2. Use the dyadic use of the ceiling and floor functions to play the following game with the computer. Toss a coin 100 times. You win \$1 whenever a head occurs, otherwise you lose \$1. You start the game with \$20. During the game record:
   (a) How many times you were in deficit.
   (b) How many times you were in surplus.
   (c) What your maximum profit was and at which toss it occurred.
   (d) What your maximum loss was and at which toss it occurred.
   (e) How many times you switched from a surplus to a deficit position.

Did you go broke by the end of the game?

3. A random variable X takes the values 1, 2, . . . , 11, each with probability of (1/11). Use the scan operator and the random number generator to confirm the following two arguments.
   (a) As the sample size gets large the sample mean "approaches" (or fluctuates more closely around) the number 6, which is the population mean.
   (b) As the sample size gets bigger the sample variance "approaches" (or fluctuates closely around) the number 10.083, which is the population variance.

4. The numbers of admissions to the emergency ward of a hospital between 4 and 8 P.M. for a period of 20 days were 0 1 3 0 2 3 4 5 0 5 3 6 3 4 0 1 1 4 2 5.
   (a) Use your *DSTAT X* function to calculate the maximum, minimum, range, mean, variance, standard deviation, mean deviation, and median.
   (b) Unexpectedly, on the 21st day there were 11 admissions. In light of this information recalculate the measures asked for in (a) and find which ones are affected (i.e., increase or decrease), and which remain unchanged.

Statistical Application Chapter 5

5. The following table gives the population growth rates for various regions of the world for the year 1970:

Region	Population in Millions	Annual Growth Rate in %
Europe	470	0.8
U.S.S.R.	240	1.1
N. America	230	1.3
Oceania	20	2.1
Asia	2100	2.3
Africa	350	2.6
S. America	290	2.9
World	3700	

Source: 1974 American Almanac, Table 1322.

   (a) Find the weighted average of the world population growth rate. Which average population (weighted, unweighted, geometric, or harmonic) is the most appropriate?
   (b) Use the computer to predict the world population for the year 2050 using the appropriate estimate for the growth rate. Compare this to the estimate obtained by adding the separate estimates for each region. Assume that the world growth rate will not change.

6. Refer to exercise 1. Verify that in no case is the mean deviation greater than the standard deviation.

7. Redefine the *DSTAT X* function as *F DSTAT1 X*, where *F* is the relative frequency of *X*. Use this function to solve the following problem.

A random variable X is defined as follows:

$$X = \begin{cases} -2 \text{ with probability } 1/3 \\ 3 \text{ with probability } 1/2 \\ 1 \text{ with probability } 1/6 \end{cases}$$

Calculate all measures of location given by the function for

(a) X

(b) X^2

(c) $2X + 3$

(d) $2X^2 + 3X + 1$

(e) Is the variance of $2X + 3$ twice the variance of X?

(f) What is the relationship (if any) between the mean value of X and the mean value of X^2?

8. A population consists of six elements with X values 10 11 12 13 14 15. One element is picked at random.

(a) What is the expected value of X?

(b) What is the expected value of X^2?

(c) Suppose that the elements are circles and the X-values are their diameters. What is the expected value of the area of such a circle?

(d) Suppose that the X values are dollars per day that you carry in your pocket during a 6-day period, and that you always spend \$4 plus 80% of whatever is left. What are your expected expenditures on a randomly selected day?

(e) Suppose that the X values are the number of cars per minute that pass through six gates at a toll road. Find the expected number of cars going through the six gates per hour.

9. Let X take the values 0, 1, 2, 3, 4, and let $Y = X^2/30$ be its probability distribution function. First verify that Y is a probability distribution function, i.e., that $Y_i > 0$ and $\Sigma_0^4 Y_i = 1$. Find $E(X)$ and the variance of X.

6

Higher and Cross Product Moments and Distributions

Higher Order Sample Moments

So far we have restricted our attention to the first two sample moments: the arithmetic mean and the sample variance. Let us write a function to calculate the rth sample moment about the mean, where r is greater than or equal to 2. Mathematically, what we want to calculate is

$$\Sigma_1^N(x_i - \bar{x})^r/(N - 1)$$

Higher Order Sample Moments

From what we have learned so far, we can write the answer down. Try doing so by working outwards from $(x_i - \bar{x})$ in the mathematical expression. Thus, write down on a piece of paper

```
First effort:          X-MEAN←(+/X)÷N
Second:             M←(+/(X-MEAN←(+/X)÷N)*R)
Third:              M←(+/(X-MEAN←(+/X)÷N)*R) ÷ (N← ρ ,X)-1
Fourth:               ∇M←R MNTS X
                [1] M←(+/(X-MEAN←(+/X)÷N)*R) ÷ (N← ρ ,X)-1
                [2] ∇
```

Now that we have it defined, let's try using it. Try

```
      3 MNTS X
22.245
      4 MNTS X
570.82
      5 MNTS X
1477.1
      6 MNTS X
20467
```

We have a function that produces an explicit result and has two arguments.

Covariance

Covariance

Another simple function we can write down is one that calculates the covariance between two arrays. The mathematical statement is

$$\text{Cov}(x, y) = \Sigma_1^N (x_i - \bar{x})(y_i - \bar{y})/N$$

$$= \Sigma_1^N x_i y_i / N - \bar{x}\bar{y}$$

where $\bar{x}$, $\bar{y}$ are the means of the arrays x and y. Try to write down the APL expressions on a piece of paper.

```
First effort:   ((X+.×Y)÷N)-((+/X)×(+/Y))÷N*2
Second:         ((X+.×Y)-((+/X)×(+/Y))÷N)÷N← ρ ,X
Third:              ∇C←X COV Y
                [1] C←((X+.×Y)-((+/X)×(+/Y))÷N)÷N← ρ ,X ∇
```

To use this new function, we will need another array, say *Y*. Type in

```
      X←1 2 ¯3 ¯4 5 ¯6 ¯7

      Y←2 4 6 8 4 2 6

      X COV Y

¯2.7347

      Y COV X

¯2.7347
```

6.1 Some Useful Distributions (Binomial, Poisson)

One of the first distributions that you would encounter in your studies of statistics would be the binomial distribution, which has associated with it the ubiquitous binomial coefficients.

Binomial Coefficients

Binomial Coefficients

Mathematically, we define the binomial coefficient by

$$\binom{n}{r} = \frac{n!}{(n-r)!\,r!}$$

for $r = 0, 1, 2, \ldots, n$, and where $n! = 1 \times 2 \times 3 \times \cdots \times n$, $r! = 1 \times 2 \times 3 \times \cdots \times r$, and $(n - r)! = 1 \times 2 \times 3 \times \cdots \times (n - r)$. The notation $n!$ is called n factorial. As you may recall, the term $\binom{n}{r}$ represents the rth term in the expansion of the polynomial $(a + b)^n$. The binomial

Binomial Probability Distribution

probability distribution is given by

$$\binom{n}{r} p^r (1 - p)^{n-r}$$

where $0 < p < 1$ is the probability of some event occurring, and $\binom{n}{r} p^r (1 - p)^{n-r}$ is the probability of getting r successes in n independent trials. $\binom{n}{r}$ is also the number of ways that n objects can be *combined* r at a time.

In APL, factorial and combinatorial functions are handled very simply. We use !, "shriek," or the exclamation point, which is keyed by upper shift K, backspace, period. Thus, !, as a monadic function, produces N factorial by executing !N. Try

Factorial Function

```
      !3
6
      !5
120
```

Now try:

```
      !0
1
      !6
720
      !!3
720
```

Combination of R Things N at a Time

The combinatorial function that we just discussed is obtained from the dyadic use of !, that is, $\binom{n}{r}$ or the number of combinations of r things taken n at a time is given by R!N. R must not exceed N, otherwise the result is zero. Try.

```
      3!5
10
      0!5
1
      5!5
1
```

Binomial Distribution

Using ! we can write a function to yield binomial probabilities as defined by $\binom{n}{r} p^r (1 - p)^{n-r}$; indeed we can just write the function out. Try the following:

Binomial Distribution

```
        ∇PR←N BI P
[1] PR← (R!N)×(P*R)×(1-P)*N-R←0,⍳N
[2] ∇
```

PR is a result array with $n + 1$ elements in it, where the rth element is the probability of r successes in n independent trials. In *BI*, *R* is computed to be an $(n + 1)$-element array with elements 0, 1, 2, . . . , n. Try

```
        5 BI .5
0.03125 0.15625 0.3125 0.3125 0.25625 0.03125
        5 BI .2
0.32768 0.4096 0.2048 0.0512 0.0064 0.00032
        5 BI .8
0.00032 0.0064 0.0512 0.2048 0.4096 0.32768
```

Probabilities are meant to sum to 1. If we have done a reasonable job of calculating these probabilities, we should be able to add them up to get 1. Let's try:

```
        +/5 BI .5
1
        +/5 BI .2
1
```

Poisson Distribution

Another important discrete distribution is the Poisson distribution with mean value M. The probability distribution is defined mathematically by

Poisson Distribution

$$\frac{e^{-M}M^X}{X!}, \qquad X = 0, 1, 2, \ldots \text{ and } e = 2.718 \ldots$$

An APL function that generates Poisson probabilities is easily written. One minor problem is that we cannot write a routine to give *all* the probabilities, because that would mean an infinite array length; even APL finds that difficult! Let us compromise and specify that we want only the first N probabilities. This suggests that we define a dyadic function. Consider

```
        ∇ PR←N POISSON M
[1] PR←(*-M)×(M*X)÷!X←0,⍳N
[2] ∇
```

X is an $(N + 1)$-length array, as is PR, which contains the probabilities of 0, 1, 2, . . . up to N successes in an infinite number of trials.

For example,

```
      5 POISSON .5
0.60653 0.30327 0.075816 0.012636 0.0015795 0.00015795
```

Cumulative Poisson Distribution

One frequently wants the cumulative probabilities, that is, the sum of the probabilities from 0 to R, $0 < R \leq N$, for the binomial distribution, or 0 to R for some integer R for the Poisson distribution. Thus, the sum of probabilities for 0 to R gives the probability of no more than R successes. Correspondingly, the probability of at least R successes is $1 - \Sigma_1^{R-1} PR_i$. Let us use our binomial probability function to generate an array of cumulative binomial probabilities. This involves the use of a primitive function that we haven't used very much, \, scan (typed by striking upper case /); see page 28 for its definition. The function we require is

```
      ∇CB←N CUMBI P
[1] CB←+\N BI P
[2] ∇
```

Note that we have a function inside a function, which is perfectly okay. This is referred to as one function calling another. In fact, a function can call itself. That technique is called recursive programming; it is an interesting subject, and Barron [1968] is a good reference. Returning to cumulative probabilities, let's try

```
      5 CUMBI .5
0.03125 0.1875 0.5 0.8125 0.96875 1
      5 CUMBI .2
0.32768 0.73728 0.94208 0.99328 0.99968 1
```

6.2 Histograms

Histograms

It is a trite saying that a picture is worth a thousand words. That doesn't stop us from repeating it, and noting that in programming it seems that to get one simple picture we need to use 10,000 words. This is not so in APL. Let us write a simple program for getting a histogram from an array of absolute frequencies. Because the number of entries in some cells may be very large, it would be useful to have a simple way of scaling the histogram down so that our plots do not take up pages and pages of output. For example, if we have a set of absolute frequencies whose sum is 100, dividing each frequency by 10 and rounding off should produce a useful histogram.* Suppose that we have an array F of absolute frequencies. Consider

```
      ∇ G←S HIST F;M;K
[1] M←⌈/F← ⌊ 0.5+F÷S
```

* This routine is adapted from K. W. Smillie [1969, p. 20].

```
[2] G←('.'),(⌈/K←(F≥M)/⍳⍴F)⍴' '
[3] G[K+1]←'T'
[4] G
[5] →(0<M←M-1)/2
[6] (1+⍴F)⍴'.'
[7] G←⍳0
[8] ∇
```

An example of the use of *HIST* is given by

```
F←3 8 10 20 9 7 4
S←2
S HIST F
.   T
.   T
.   T
.   T
.   T
.  TTT
. TTTTT
. TTTTT
.TTTTTTT
.TTTTTTT
........
```

You will notice that the function starts by plotting a *T* in the position of the largest frequency first, and then moves *down* toward the lower frequencies.

In the first line, the operation `⌊0.5+F÷S` divides each frequency by *S*, adds 0.5 and rounds down to the nearest integer. In short, this procedure correctly produces the usual rounding-off of numbers to the nearest integer after division of *F* by the scale factor, *S*. If *F*/*S* is 4.2 or 4.5, `⌊0.5+F÷S` produces 4 or 5, respectively. The rounded numbers are restored in *F*.

Dyadic Functions Maximum and Minimum

Dyadic Functions Maximum, ⌈ and Minimum, ⌊

The dyadic operation `⌈/F` finds the maximum of the array *F* and stores the result in *M*. `⌈/F` is equivalent to

$f_1 \lceil f_2 \lceil f_3 \ldots$

and the *dyadic* use of the symbol ⌈, called maximum, is to produce the larger number of each pair compared. Thus

```
3⌈5
5
```

```
      5⌈3
5
      Y
2 4 6 8 4 2 6
      ⌈/Y
8
```

The symbol ⌊, in its dyadic mode, is the minimum function. So line [1] in *HIST* produces an array F of scaled and integer-rounded frequencies and stores the largest frequency in M.

Dyadic Function Reshape

Line [2] of *HIST* introduces the dyadic use of ρ, called reshape. It is keyed by striking upper shift *R*. The operation *N*ρ*V* will make V into an array of length N. If V is too big (too many elements), then only the first N elements of V will be used, and if V is too small, the elements of V will be repeated in sequence until the newly created array is of length N. Some examples of this function are

Dyadic Reshape ρ

```
      2 ρ Y
2 4
      7 ρ Y
2 4 6 8 4 2 6
      9 ρ Y
2 4 6 8 4 2 6 2 4
```

Nine is greater than the number of elements in Y, so the two extra elements came from the beginning of Y.

```
      4 ρ 3
3 3 3 3
      4 ρ 'A'
AAAA
```

Keep this result in mind for the next paragraph.

```
      1 ρ 2
2
      ρ 2

      ρ A←1 ρ 2
1
```

Compare these last three examples carefully.

Character Arrays

Line [2] of *HIST* uses the symbols ' ' (upper shift *K*). The use of a pair of quotation marks, as you probably remember, tells the computer to regard whatever is between them as an array of "literals," that is, characters that are not to be executed. Thus we get

```
      4 ρ '.'
....
      3 ρ 'T'
TTT
      3 ρ ' '
& & &
```

and so on.

Logical Functions

Logical Functions <, ≤, =, ≥, >, ≠, ∧, ⍲, ∨, ⍱

In line [2] of *HIST* the left-hand argument to the dyadic function ρ is (⌈/K←(F≥M)/ιρF). The result is an integer, although at the moment it all appears to be highly mysterious. The operation ιρF, you may recall, produces the array of index numbers 1,2,...,ρF, where ρF is the number of frequencies.

The expression (F≥M) introduces a new type of primitive function, the so-called logical functions. Other examples of logical functions are: >, <, ≤, =, ≠, etc. Essentially, the logical function asks a question, say, Is the relationship $a < b$ true or false? If true, return a one, if false return a zero. For example,

```
      3 < 1
0
      5 < 3
0
      4 ≥ 4
1
      2=1+1
1
      2≠3-1
0
```

You see that the output of a logical function, the result if you like of a logical comparison, is either 0 (false) or 1 (true). In our example, F is an array and M is the maximum value of the elements in the array F. If F has only one maximum, say in the Ith position, then $(F \geq M)$ produces an array of the same length as F with 0's everywhere except in the Ith position, which contains a 1. That is, $(F \geq{\cdot}M)$ looks like 0 0 0 1 0 0. Try, for example

```
      3<1 2 3 4 5
0 0 0 1 1
      10>1 2 3 4
1 1 1 1
```

A second group of logical functions not only produces results that are either `1`'s or `0`'s but also requires that the arguments of the function be `1`'s and `0`'s. These logical functions are: 'or' ∨, 'and' ∧, and 'not' ~. To see how these functions work look at the Truth Table for FfM

Truth Table

F	M	∨	∧	⍱	⍲
1	0	1	0	0	1
1	1	1	1	0	0
0	0	0	0	1	1
0	1	1	0	0	1

The table gets its name from the fact that a statement like, "the mean of x is 22.3 and x has a variance of 2.3" is true if and only if both parts of the conjunction are true. We could have a long series of ∧ (and's) and for the statement to be true (result in a 1) every element would have to be true (be a 1). The statement, "the mean of x is 22.3 or the variance of x is 22.3," is true if either part is true. We use the symbol ~ (not) to change a 1 to a 0, or a 0 to a 1. And if ∨ or ∧ are overstruck with ~ the results are "negated."

These logical functions have many uses outside of symbolic logic. For example, in modeling material or traffic flow you may have a process that can be diagrammed as

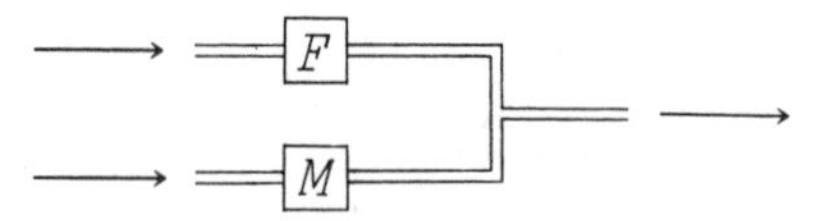

Here to stop the process you need to have both F and M stopped. To obtain flow, either one needs to be on. If the system looked like

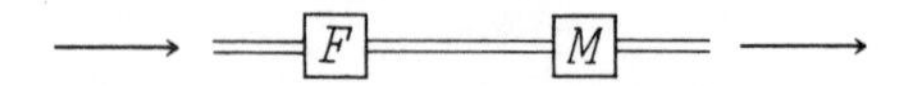

Logic Functions in Flow Systems

both F and M would have to be on to obtain production or flow and either could be off to stop it. You can see the direct relation between these situations and our logical functions. After we show you the compress function we will show you some numerical applications of these ideas.

Compression

Compression /

We can now consider the dyadic use of /, which is called compression. The left argument must be composed of 1's and 0's only. Both arguments must be arrays of the same length, except that the left-hand side may be a scalar. What happens is this. Given an array, say A, of 0's and 1's on the left-hand side, and an array, say B, of equal length on the right, then A/B produces an array whose length is equal to the number of 1's in A. Whenever a_i is 0, b_i is dropped, and whenever a_i is 1, b_i is retained. For example,

```
        A←0 0 1 0 1 0 0 1
        B←1 2 3 4 5 6 7 8
        A/B
3 5 8
        1/B
1 2 3 4 5 6 7 8
        0/B
```

The blank space after $0/B$ means that the 0 compression of B is an empty vector which, when you consider the matter, is natural enough.

Back to the expression `(⌈/K←(F≥M)/⍳⍴F)`. What is put into K is the *array of index numbers* where F has maximum frequency. `⌈/K` picks out the *largest* of these index numbers. This determines the number of blank spaces to be catenated to the symbol `'.'`. This completes line `[2]` of *HIST*.

Line `[3]` gives rise to no difficulties. K is an array (even if it has only one element), of which each element is an index number of the array F which has maximal value. What line [3] does is to put the character T, or the literal 'T' in each position indicated by the index numbers $K + 1$, recognizing that the first element of G is `'.'`. Line `[4]` merely prints G.

Line `[5]` decides whether to continue to line `[6]` or go back to line `[2]`. What this routine does is to start at the *top* of the histogram and work its way down to the bottom, so line `[5]` first reduces M by 1. If the value of M is greater than 0, then 1/2 gives 2 and `→2` means go to line `[2]`. If M reduced by 1 is less than 0, then 0/2 gives an *empty array*. The computer interprets such a statement as one to be ignored, so it continues to the next line.

Every time M is reduced, more frequencies become eligible to be bigger than M, so more T's will be printed.

Line `[6]`, reached when M has been reduced to ¯1, merely prints a row of periods across the bottom of the graph. Line `[7]` makes G an empty array and the routine is ended. The reason for this is that if G were not redefined as an empty array, the completion of the function would print the contents of G, since an explicit result is specified in the function header.

You might note that this routine has three dummy variables—G, S, and F; two local variables—M and K; and no global variables. Try the following examples:

```
      S←10

      F←1 6 28 42 23

      S HIST F
:     T
:    TT
:    TTT
:   TTTT
:........

      F←0 0 0 0 0 1 0 1 5 10 17 21 20 15 7 3 0

      S←10

      S HIST F
:          TTTT
:        TTTTTTT
:..............°°°°

      S←1
      S HIST F
:            T
:            TT
:            TT
:            TT
:           TTT
:           TTT
:           TTTT
:           TTTT
:           TTTT
:           TTTT
:           TTTT
:          TTTTT
:          TTTTT
:          TTTTT
:          TTTTTT
:          TTTTTT
:         TTTTTTT
:         TTTTTTT
:         TTTTTTTT
:         TTTTTTTT
:     T TTTTTTTTT
:..................
```

In some cases you may want to look at the histogram of a range of your data, maybe the central portion, or possibly select the observations on a series of criteria. We might have a vector *INC* containing a sample of personal incomes and only want to look at those greater than $20,000 and less than or equal to $50,000. The APL statement would be

```
((X>20000)∧(X≤50000))/X
```

Dyadic Compression

Here we are combining the two types of logical variables and using dyadic compression. Each operation in parentheses produces a logical vector with the shape ρX. After these three functions are computed, compression selects particular elements of X that are both larger than 20,000 and equal to or smaller than 50,000. This statement could have had an assignment to a new variable F or it could have been written as

```
S HIST ((X>20000) ∧ (X≤50000))/X
```

Another situation arises in presenting statistical results and in computation in general; and that is selection of data from a larger data base. For example, along with income we may have a second vector coded with years of education, a third with degree status. Suppose we coded degree as 0-no, 1-high school, 2-associate, 3-bachelors, 4-masters, 5-Ph.D., 6-DDS, 7-MD, etc. Now suppose you want the income histogram of people with 16 or more years of formal schooling or having a 2 or 4 year degree. Suppose the data were organized on an individual-by-individual basis:

```
    INC
15843      21842      9823      13586      ...
    ED
  15         18        12         13       ...
    DEG
   2          5         0          4       ...
```

The first subject has an income of $15,843, went to school for 15 years, and has an associate of arts degree. The APL statement that selects our sample is

```
((ED≥16) ∨ (DEG=2) ∨ (DEG=1))/INC
```

You can select a number of complex combinations of attributes for analysis using these techniques.

6.3 The Normal Distribution*

Normal Distribution

The most important distribution you will have to handle in statistics is the *Normal Distribution*. The mathematical expressions for the normal density function $f(x)$ and the corresponding cumulative distribution function $F(x)$ are:

$$f(x) = \frac{\text{Exp}\,(-1/2(x-\mu)^2/\sigma^2)}{(2\pi\sigma^2)^{1/2}}$$

$$F(x) = \int_{-\infty}^{x} \frac{\text{Exp}\,(-1/2(x-\mu)^2/\sigma^2)}{(2\pi\sigma^2)^{1/2}}\,\text{dx},$$

where Exp $(\cdot)$ denotes the exponential function, i.e., $\text{Exp}(x) = e^x$.

* Remember that, as mentioned in the introduction, starred sections involve statistical material beyond the level normally presented in an elementary course. No new APL expressions, functions, or procedures which will be used later in the book will be introduced in these sections.

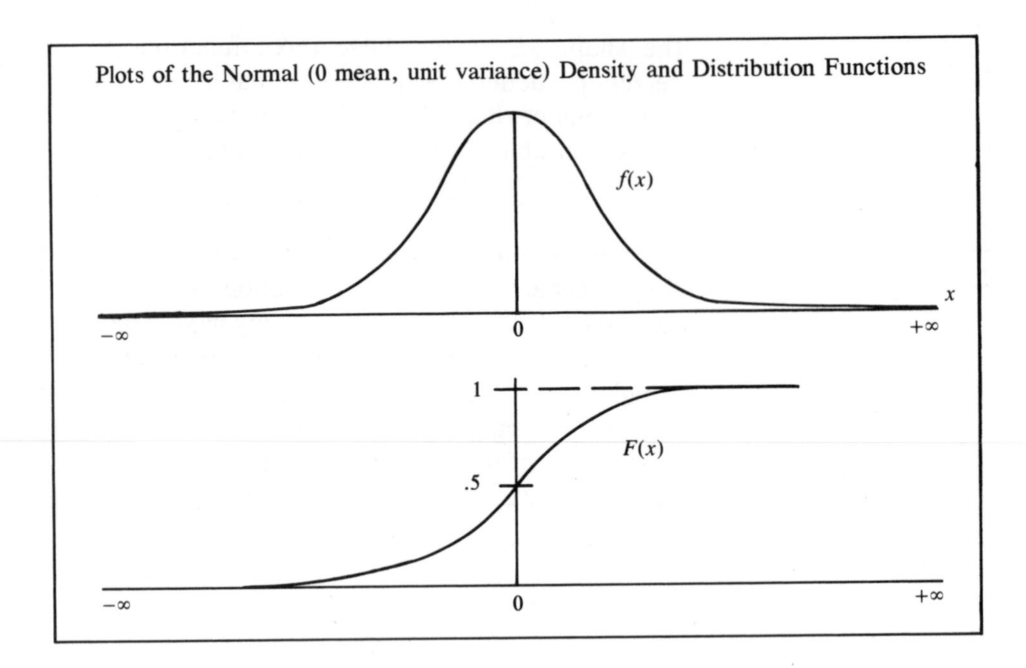

Figure 6.1

Let us define APL functions to give us the normal density and cumulative distribution values for any value of x and any value for the mean, μ, and the variance, σ^2.

In defining our functions, let us agree to let the array P have two elements, the first of which is the mean and the second the variance. Let the second argument be the x value at which we want to evaluate the functions. Thus,

"Standard"-Normal Density Function

```
P←0 1
X←3
```

means that we want to evaluate a normal density function with mean 0, variance 1, at the point x = 3. We define:

```
    ∇D←P NORMD X
[1] D←(*(-0.5×(X-P[1])*2)÷P[2])÷(○P[2]×2)*0.5
[2] ∇
```

Pi Times, ○

The only operation in this function with which you will not be familiar is the monadic function "pi times" (○ called large circle). ○ is keyed as upper case *O*. ○3 produces $\pi \times 3$ or 9.4244777961, where $\pi = 3.141592654$. ○P[2]×2 produces the mathematical expression ($2\pi\sigma^2$), where σ^2 is the variance.

Experimenting, we obtain

```
P←0 1
P NORMD 1
```

```
0.24197
        P NORMD 2
0.053991
        P NORMD ¯1
0.24197
```

Now let us get the corresponding cumulative distribution, or at least a reasonable approximation to it.

The integral of the function `P NORMD X` can be approximated by adding the areas of a series of small rectangles which approximate the area under the density function. Let the base of each approximating rectangle have a width specified by the user, and let the height be determined by the function `NORMD` at the midpoint of the interval.

Since the normal integral is theoretically defined from $-\infty$, to $+\infty$, we will have to "approximate" the end points.

Let the two arguments of the cumulative distribution function be X and the array I. The first element of I is the mean, the second the variance, the third the chosen interval width in terms of standard deviations, and the fourth the number of standard deviations below the mean at which integration starts, e.g.,

```
I←0 1 1 6
```

indicates that the distribution has a mean of 0, a variance of 1, an interval width of 1 standard deviation, and the integration begins at $(0 - 6) = -6$ on the X axis, or 6 standard deviations to the left of the origin. We define (in several lines for clarity):

```
    ∇ A←I NORMC X
[1] LHS←I[1]-I[4]×S←I[2]*0.5
[2] X←(X-I[1])÷S
[3] NINT←⌊(X-LHS)÷I[3]
[4] A←I[3]+.×I[1 2] NORMD (( LHS+I[3]×⍳NINT)-I[3]÷2)
[5] ∇
```

The first line determines `LHS`, the lefthand side from which the integration is to start (in our example ¯6). Line `[2]` normalizes the variable of integration, i.e., subtracts the mean from the point where we are evaluating the function and divides the result by the standard deviation. The third line defines `NINT` as the number of intervals into which the integral is to be broken up. The last line approximates the area under the normal density. The density function is evaluated at the midpoint of each interval.*

The procedure is illustrated in Figure 6.2.

* Note that the righthand argument in this function, X, may only be a scalar as the function is currently defined, since if X is an array, attempted execution gives a RANK ERROR. Note also that this function is valid only if the mean is zero. The routine assumes the variance is one. Also, the routine will bomb if you ask for X outside of the LHS.

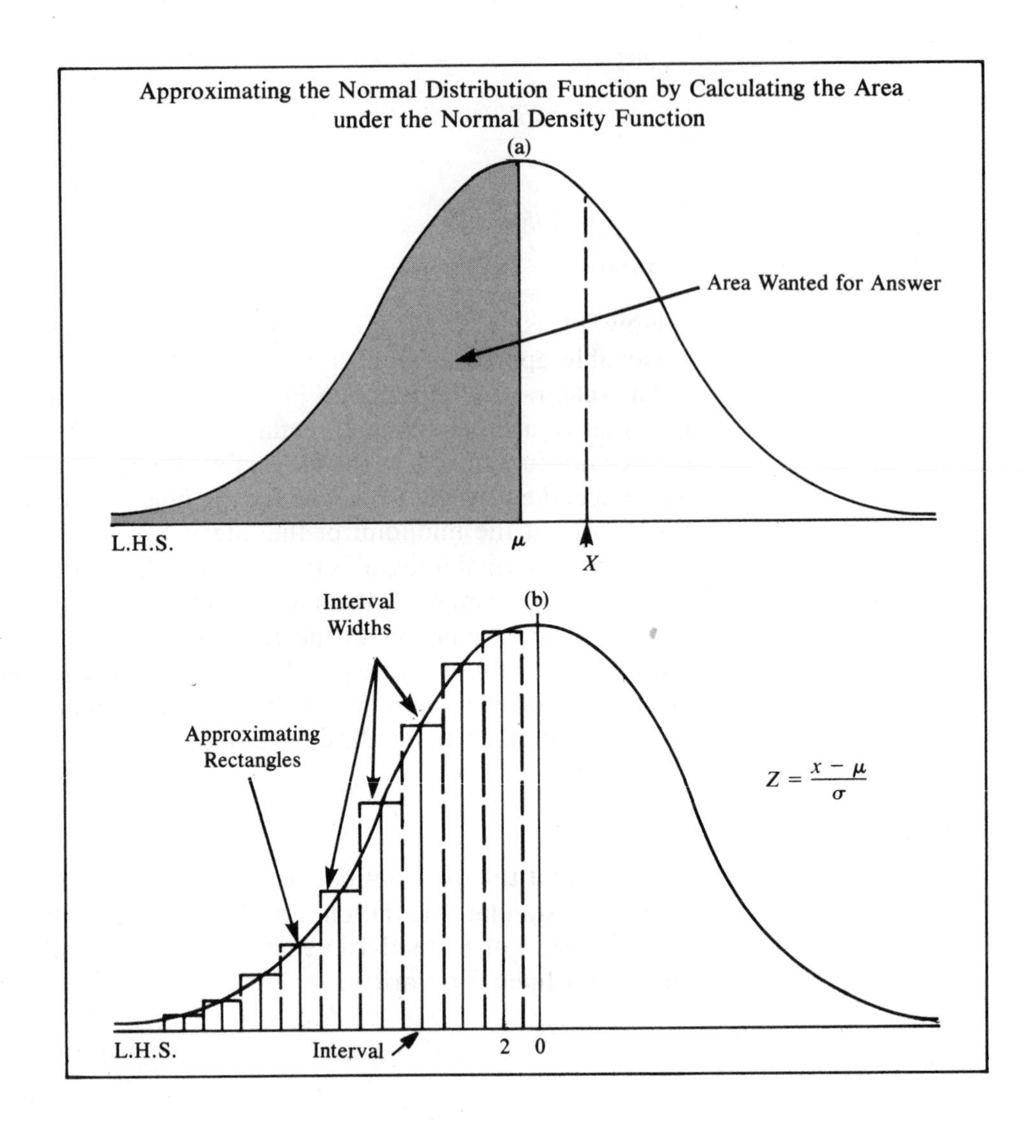

Figure 6.2

Let us check our function against a set of normal tables. Try

```
      I←0 1 .1 5
      X←¯1
      I NORMC X
0.15855
      X←0
      I NORMC X
0.5
      X←2.5
      I NORMC X
```

```
0.99381
      I NORMC 3.3
0.99952
```

Comparison of APL-STAT Routine to Tabulated Values

You can see from these results that the routines are quite accurate, since for most purposes only three significant digits is enough. Better approximation could be obtained by using a smaller value for $I[3]$ and a larger value for $I[4]$.

Biometrika Tables*

Ordinate Value	Density		Cumulative Distribution	
x	Computed	Table	Computed	Table
¯1	0.24197	0.24197	0.15855	0.15866
0	0.39894	0.39894	0.5	0.5
1	0.24197	0.24197	0.84145	0.84134
2	0.953991	0.95399	0.97729	0.97725
2.5	0.017528	0.017528	0.99381	0.99379
3.3	0.001723	0.001723	0.99952	0.99952

* Values obtained from the Pearson, E. S. & Hartley, H. O., *Biometrika Tables for Statisticians,* Vol 1. (Cambridge University Press, 1962, p. 104).

Summary

Higher Order Sample Moments are defined mathematically by

$$\Sigma_1^N (x_i - \bar{x})^r/(N - 1) \qquad \text{for } r = 2, 3, \ldots$$

APL routines to calculate them are given on page 72.

Sample Covariance is defined mathematically by

$$\begin{aligned} \mathrm{Cov}(X,Y) &= \Sigma_1^N (x_i - \bar{x})\,(y_i - \bar{y})/N \\ &= \Sigma_1^N x_i y_i/N - \bar{x}\bar{y} \end{aligned}$$

An APL routine to calculate it is given on page 73.

Binomial Coefficient:

$$\binom{n}{r} = \frac{n!}{(n - r)!\,r!}$$

where $n\,!$ is defined by $n\,! = 1 \times 2 \times 3 \times 4 \times \cdots \times n$.

Factorial, ! (of an integer), (keyed by upper shift *K*, backspace, period): `!N` produces the product `×/⍳N`

Combinatorial, !, dyadic use of previous function: `R!N` produces the binomial coefficient $\binom{n}{r}$ defined above.

The binomial probability distribution is defined by

$$\binom{n}{r} p^r (1 - p)^{n-r}$$

for r successes in n trials, where probability of a success is p. An APL function to calculate it is given on page 75.

The Poisson probability distribution is defined by

$$(e^{-M}M^X)/X!, \qquad X = 0, 1, 2, \ldots, \text{ and } e = 2.718\ldots$$

An APL function that calculates the probabilities is given on page 75.

Maximum, ⌈, (upper shift *S*): dyadic function which picks the larger of its arguments.

Minimum, ⌊, (upper shift *D*): dyadic function which picks the smaller of its arguments.

Reshape, ⍴, (upper shift *R*): dyadic function which rearranges righthand argument into an array with the number of elements determined by the lefthand argument.

Quotes ' ', (upper shift *K*): entries between quote symbols are treated as characters, or literals, not as digits and numbers, or variable names, or APL functions.

Logical functions (<, ≤, >, ≥, =, ≠): dyadic functions which compare left and right arguments. If the stated relation is true for the arguments compared, a 1 is produced by the function; otherwise a 0 is produced. Examples:

```
      3<4
1
      4<3
0
```

Logical functions (∧, ⍲, ∨, ⍱): dyadic functions which produce binary results as indicated in the truth table on page 80.

Compression, /: dyadic use of reduction. Left argument must be either 0, 1, or an array of 0's, 1's equal in length to righthand array variable. Output is an array of length equal to the number of 1's and whose elements are the elements of the righthand array selected according to the position of the 1's.

EXAMPLES:

```
      1/1 2 3
1 2 3
      1 0 1/1 2 3
1 3
      0/1 2 3
      0 1 0/1 2 3
2
```

See pages 76 and 77 for an APL function that plots histograms.

In the starred section, the normal density and cumulative distribution functions are calculated with APL routines.

Exercises

APL Practice

1. Let's explore some of the uses of the logical functions. Let W←?20ρ5 and K←'AN APPLE A DAY KEEPS THE DOCTOR AWAY'.

   (a) W=5

   (b) W < 14

   (c) W≠1

   (d) W≥7

   (e) 'A'=K

   (f) 'P'=K

   (g) 0 > W

   (h) W≤0

   (i) 'K'≠K

   (j) +/'E'=K How many E's do we have in K?

   (k) +/W=3 How many 3's do we have in W?

   (l) Find the frequency of 2 in W.

   (m) F←(+/W=1), (+/W=2), (+/W=3), (+/W=4), (+/W=5), the row of absolute frequencies.

   (n) W='K'

   (o) W=K

   (p) X-.=3; the same result in (K).

   (q) +/W+.≤3; the number of elements of W that are smaller than or equal to 3.

   (r) FR←F÷ρW; a list of relative frequencies.

   (s) +\FR; the cumulative distribution of W where F is defined in (m).

2. The Maclaurin's series for the cosine of X in *radians* is given by

$$\cos X = \frac{X^0}{0!} - \frac{X^2}{2!} + \frac{X^4}{4!} - \frac{X^6}{6!} + \frac{X^8}{8!} - \cdots$$

Use the relations

$$\sin X = \sqrt{1 - \cos^2 X}$$

$$\tan X = \frac{\sin X}{\cos X}$$

$$\operatorname{cotan} X = \frac{1}{\tan X}$$

to find the sine, cosine, tangent and cotangent of 30°. (*Hint:* use only the first twenty terms of the series. 1 degree = 3.14/180 radians)

3. Show that

   (a) $\Sigma_0^N (-1)^j \binom{N}{j} = 0$ for $N = 50$, $j = 0, 1, \ldots, 50$.

   (b) $\Sigma_0^N \binom{N}{j} = 2$ for $N = 50$, $j = 0, 1, \ldots, 50$.

(c) $\Sigma_1^N j\binom{N}{j} = N \cdot 2^{N-1}$ for $N = 50$, $j = 1, 2, \ldots, 50$.

(d) $\Sigma_1^N (-1)^{j-1} j\binom{N}{j} = 0$ for $N = 50$, $j = 1, \ldots, 50$.

(e) $\binom{N}{r} = \Sigma_0^r \binom{K}{j}\binom{N-K}{r-j}$ for $j = 0, 1, \ldots, 20, r = 20, K = 30, N = 60$.

(f) $\binom{2N}{N} = \Sigma_0^N \binom{N}{j}^2$ for $N = 50$, $j = 0, 1, \ldots, 50$.

(g) $\Sigma_1^{10} L^j e^{-L}/j! \doteq 1$ for $L = 5, j = 1, \ldots, 10$.
where $\doteq$ means "approximately equal."

(h) $\binom{N}{K} = \binom{N-1}{K} + \binom{N-1}{K-1}$

(i) $\binom{N}{K}\binom{r-N}{M-K}/\binom{r}{M} = \binom{M}{K}\binom{r-M}{N-K}/\binom{r}{M}$
for $r = 25$, $M = 20$, $N = 15$, $K = 10$.

4. An interesting fact is that you can express the relational logical functions <, ≤, =, ≠, >, ≥, in terms of ~, ∨, ⍱, ∧, ⍲, when the arguments are binary variables. For example, when A and B are either 1 or 0, `A>B` can be expressed as `A∧~B`. In order to see these relationships more clearly, write all of the relational logical functions in terms of ~, ∧, and ∨.

* **5.** Evaluate the integral

$$F(x) = \int_0^3 x^3 dx, \text{ using } \texttt{X←(⍳30)÷10}$$

6. In general, for any two arrays A and B with equal numbers of elements, and any two binary functions "f" and "g", the expression $Af \cdot gB$ (the inner product of A and B) produces a scalar of the following form: $(A_1 g B_1) f (A_2 g B_2) f (A_3 g B_3) f \cdots$. For $A \leftarrow 2\ 3\ 5\ 6$ and $B \leftarrow 4\ 5\ 5\ 6$, try the following:

(a) `A+.×B`
(b) `A×.+B`
(c) `A+.=B`
(d) `A=.+B`
(e) `A+.!B`
(f) `A=.=B`
(g) `A-.=B`

7. Consider the function $f(X) = 10X - X^2$ where `X←0,⍳10`

(a) First find the integral $\int_0^{10} f(X)dX$ algebraically.

(b) Write a function that will calculate the same integral utilizing Simson's Rule which is to be explained below: Simson's Rule divides the interval (0, 10) into an *even* number of N subintervals, namely, $\delta_1 X$, $\delta_2 X$, . . . such that $\delta_1 X = X_1 - X_0 = X_2 - X_1$, $\delta_3 X = X_3 - X_2 = X_4 - X_3 \ldots$ The area under the curve will be given by the sum

$$\delta_1 X \frac{f(X_0) + 4f(X_1) + f(X_2)}{3} + \delta_3 X \frac{f(X_2) + 4f(X_3) + f(X_4)}{3}$$
$$+ \delta_5 X \frac{f(X_4) + 4f(X_5) + f(X_6)}{3} + \cdots$$

(c) Write a routine that will calculate the same integral utilizing the Trapezoidal Rule.

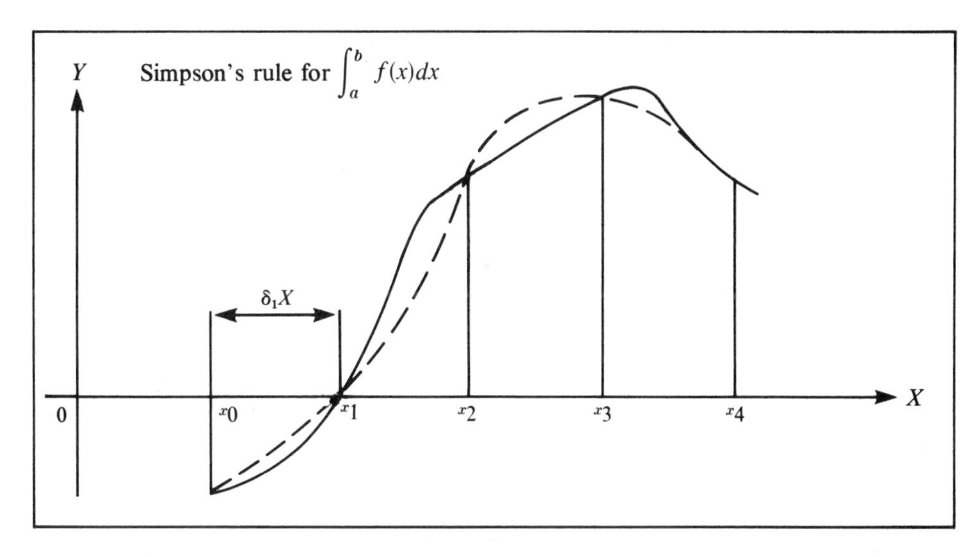

Figure 6.3

Trapezoidal Rule: Divide (0, 10) into N equal intervals $\delta_1 X$, $\delta_2 X$, $\delta_3 X$, . . . Use the formula

$$\delta_1 X \frac{f(X_0) + f(X_1)}{2} + \delta_2 X \frac{f(X_1) + f(X_2)}{2} + \delta_3 X \frac{f(X_2) + f(X_3)}{2} + \cdots$$

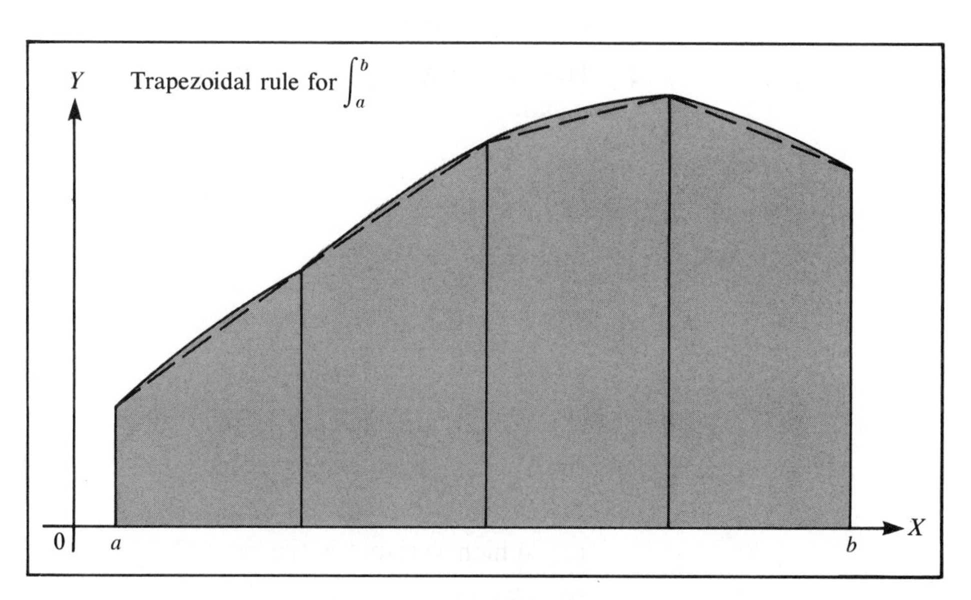

Figure 6.4

(d) And lastly, use the usual rectangular rule, discussed in the text, to calculate the integral. The rectangular rule uses the formula

$$\delta_1 X \cdot f(X_1) + \delta_2 X \cdot f(X_2) + \delta_3 X \cdot f(X_3) + \cdots$$

(e) Which method gives you the most accurate result? which the least?

(f) Which method is more appropriate for
   (a) Convex functions?
   (b) Concave functions?

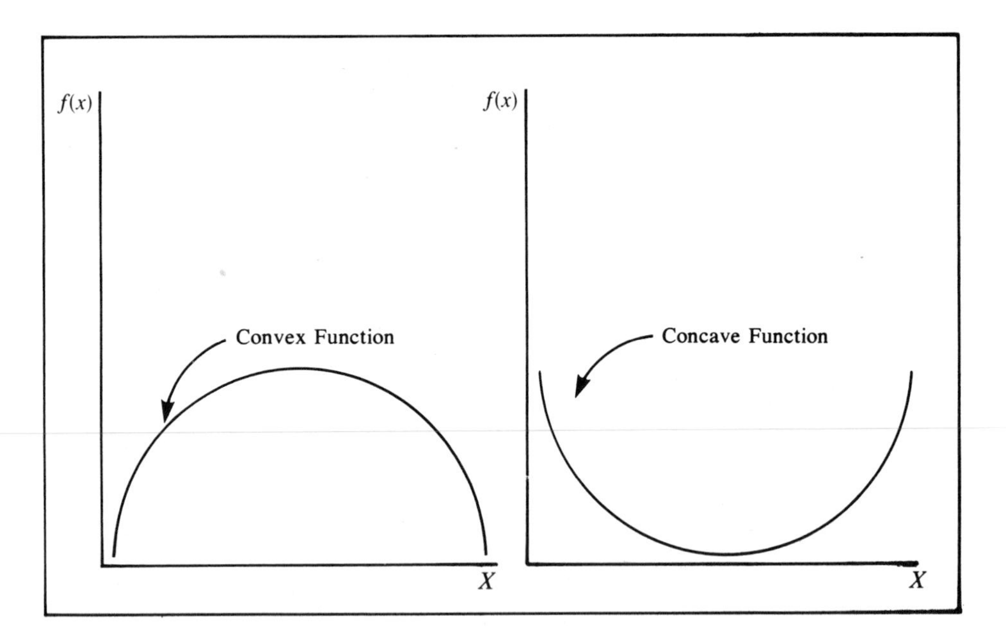

Figure 6.5 Illustration of a Concave and a Convex Function

Statistical Applications

1. Here is a function that gives you the Rth sample moment about the mean.

```
     ∇ M←R MNT  X;N;M
[1] N←⍴ ,X
[2] M←(+/X)÷N
[3] M←+/((X-M)*R)÷N-1
[4] ∇
```

(a) Which variables are local?
(b) Which variables are global?
(c) Which variables are dummy?
(d) Which variable is the result?
(e) Define $X \leftarrow$ ¯3 ¯2 ¯2 ¯1 ¯1 ¯1 0 0 0 0 1 1 1 2 2 3 and find the first four moments of X.
(f) Delete the first ¯1 from X to get a list of length 15 and repeat (d).
(g) Compare (d) and (e).
(h) Given that the first four moments of $N(0, 1)$, a normal distribution with zero mean and unit variance, are 0, 1, 0, 3, compare the results in (d) with the population moments of an $N(0, 1)$ variable.

2. Use the binomial function (page 75) to solve the following problem. Thirty percent of the people in Barrington, Illinois, have blue eyes. In a random sample of 10 find the

(a) probability that exactly 5 have blue eyes.

(b) probability that no more than 5 have blue eyes.

(c) probability that fewer than 5 have blue eyes.

(d) probability that at least 5 have blue eyes.

(e) probability that no fewer than 5 have blue eyes.

3. Let the variable X take the values $\pm 1/4$, $\pm 1/2$, ± 1 with equal probability, and let $Y = X^2$. Verify that $\text{cov}(X, Y) = 0$, i.e., X and Y are not correlated, though they are obviously dependent since $Y = X^2$.

4. The probability of getting the first success on the kth trial is given by $f(k) = p(1 - p)^{k-1}$ (geometric distribution) where p is the probability of a success in single trial. If the probabilities of having a male and a female offspring are equal, find the probability that a family's fourth child is their first son.

⋆ **5.** Let $f(X) = e^{-X}$, $X > 0$, be a probability function. Use the terminal to search for a value of X, call it X_0, such that $\int_0^{X_0} e^{-X} dX = 1/2$, i.e., the probability that $X < X_0$ is equal to 50%. (*Hint:* use $X \leftarrow (\iota 100) \div 100$, the scan operator, and the rectangular rule described in exercise 7 (d), of this chapter.)

⋆ **6.** Use the Poisson and the binomial distributions (page 75) to verify the following statement, "The one distribution approximates the other whenever the sample size is very large and the probability of success is small." In light of this statement find the probability that exactly two out of a 100 full-time traveling salesmen will be involved in a serious automobile accident during a year if the probability of one driver being involved in an auto accident during a year is .01. (See exercises 7 and 8.)

7. Write a function, name *BI TABLE,* that will construct the binomial tables for $N = 2, 3, \ldots, 10$, $R = 0, 1, \ldots, N$, and $P = 0.05$, 0.1, 0.2, 0.3, 0.4, 0.5, 0.6, 0.7, 0.8, 0.9, 0.95. (*Hint:* use the *N BI P* function (page 75) and a conditional branching that will repeat the same step for different N's and P's.)

8. Write a function, called *POISSON TABLE,* that will construct a table of the Poisson distribution probabilities for $M = 0.01, 0.02, \ldots, 1, \ldots, 9.9, 10$; $N = 0, \ldots, 10$. (*Hint:* use *N POISSON M*, page 75, and the conditional branching to execute *N POISSON M* as many times as needed.) Note: if the sample size is very large, e.g., 200, and the probability of a success very small, e.g., $P = 0.03$, you cannot use the binomial because it involves a lot of calculations. Instead, you can use the Poisson with $M = NP$, e.g., $M = (100)(.01) = 1$, and N depending on the problem.

9. Write a function that will solve the following types of problems.

In a food processing and packaging plant there are on the average two packaging machine breakdowns per week. Assume that the weekly machine breakdowns follow a Poisson distribution and find

(a) The probability that there are no machine breakdowns in a given week.

(b) The probability that there are no more than two machine breakdowns in a given week.

(c) The probability that there is at least one breakdown in a given week.

(d) The probability that there is at most one breakdown in a given week.

10. The sample coefficient of skewness is defined as $a_3 = M_3/S^3$, where M_3 is the third moment about the mean and S is the standard deviation. Note:

if $a_3 > 0$ the distribution is skewed to the right.
if $a_3 < 0$ the distribution is skewed to the left.
if $a_3 = 0$ the distribution is symmetric.

Find the coefficient of skewness of X in exercises 1 (d) and 1 (e).

11. The coefficient of kurtosis measures the "sharpness" with which a distribution peaks. It is given by $a_4 = M_4/S^4$, where M_4 is the fourth moment about the mean and S is the standard deviation. Find the sample coefficient of kurtosis for the list X given in exercises 1 (d) and 1 (e).

12. Let `X←0,⍳10` be binomially distributed, with `P=0.5` and `N=10`. Use the binomial function (page 75) to find the sample probability distribution and compare this distribution to the normal with respect to skewness and kurtosis.

13. Repeat exercise 12 with `X←0,⍳10`, $P = 4$, and $N = 10$.

14. Repeat exercise 12 with `X←0,⍳10`, $P = 0.6$, and $N = 10$.

15. The joint probability distribution function of two discrete random variables X and Y is

$$f(x, y) = \begin{cases} cxy & \text{if } 0 \le x \le 4, \quad 1 \le y \le 5 \\ 0 & \text{otherwise} \end{cases}$$

where $x = 0, .1, .2, \ldots, 3.0, \ldots, 3.9, 4.0$ and $y = 1, 1.1, 1.2, \ldots, 3.0, \ldots, 4.9, 5.0$.

Let `X←(0,⍳40)÷10` and `Y←(Y≥1)/Y←(0,⍳50)÷10`

(a) Find c such that $f(x, y)$ is a probability distribution function, i.e., $\Sigma_1^{41}\Sigma_1^{41} cx_iy_j = 1$.

(b) Find the joint probability $P(1 \le x \le 3, 2 \le y \le 3)$.

(c) Find the joint probability $P(x \ge 3, y \le 2)$.

(d) Find the $f(x; y = 2)$, i.e., the conditional probabilities of x when $y = 2$.

(e) Find the conditional expectation $E(x; y = 2)$, i.e., find the conditional mean value of x given $y = 2$.

(f) Let $Z \leftarrow 2 + (3 \times X) + 4 \times X * 2$, $(Z = 2 + 3X + 4X^2)$. Find $E(Z; Y = 2)$, i.e., the mean value of Z given $Y = 2$.

(g) Find $E[(X - E(X))^2, Y = 2]$, i.e., the variance of X given $Y = 2$.

(h) Find the standard deviation of X given $Y = 2$, i.e., the square root of your answer to (f).

16. Refer to exercise 15 and note that for a given value of Y we get a vector of probabilities of X hence the corresponding values for $E(X)$ and the variance of X. Find the variance of X for $Y = 2$ and $Y = 3$. Are the two values the same?

17. The following table shows the number of days in a 50-day period on which X automobile accidents occurred in a city.

Number of Accidents	Number of Days f
0	30
1	14
2	3
3	2
4	1
5	0
6	0
	50

Compute:

(a) The mean value of the sample ($\bar{X}$).

(b) The variance.

(c) The standard deviation.

(d) The coefficient of skewness.

(e) The coefficient of kurtosis.

(f) Assume that we want to compare this sample distribution to the theoretical Poisson distribution with Y = prob (X accidents) = $M^X e^{-M}/X!$, where we replace M, the mean value of the theoretical distribution, with the sample mean $\bar{X}$. Find Y for $X = 0, 1, 2, 3, 4, 5, 6$. Find the theoretical number of accidents (i.e., 50 times Y).

(g) The theoretical variance is equal to M (which in this case is estimated by $\bar{X}$). Is the sample variance equal to $\bar{X}$?

(h) Why might we wish to compare the sample distribution of accidents with the theoretical Poisson distribution?

18. Put the *HIST* function (page 76) into your workspace. Let `F←(⍳35)`, `(⌽⍳35)`, and then try

(a)
```
 1  HIST  F

 2  HIST  F

 3  HIST  F

10  HIST  F
```

What is the difference between these 4 histograms?

(b) Draw the histogram of the number of accidents vs. the number of days from exercise 17.

7

Data and Information— How to Get It In and Out

7.1 Numeric and Character Arrays

We begin this chapter with a brief review. In Chapter 2 you learned how to store numeric data in a variable name. You used the left-pointing arrow or the specification function, ←; for example, typing A←1,2,3,4 followed by EXECUTE or RETURN stores this array of four numbers in locations which can be referenced when needed by typing A. In APL you can separate the numbers by spaces (blanks), as we have been doing, or by commas. Thus you can specify the array by

```
      A←1, 2,3,4
      A
1 2 3 4
```

Review Numeric Arrays

This specification is made possible by using the comma (,) which plays the role of the (primitive) catenate function (see pages 41–42).

Next, suppose you want to add more data to the *A* array, for example, data contained in array *B*.

```
      B ← 5 6 7 8 9 10
      A ← A, B
      A
1 2 3 4 5 6 7 8 9 10
```

You have joined two vectors by using the catenate function. Rather than joining the whole vector, you might want to catenate some elements—say the first, third, and sixth elements. You could type

```
      A ← A, B[1, 3, 6]
      A
1 2 3 4 5 6 7 8 9 10 5 7 10
```

A ← A,B[1&3&6], would give the same result, and is more usual and perhaps quicker. Remember that our *last* definition of *A* was A ← A, B, so the above operation is equivalent to A←A,B,B[1,3,6]. You can display elements of an array in the same way. For example:

```
      A[1 3 10 4]
1 3 10 4
```

or

```
      B[6 6 2 1 1]
10 10 6 5 5
```

So far we have stored and retrieved numerical data. We can store character information with the same commands. To store the alphabetic string *ECONOMETRICS* in vector C, we use the function quote ' ' (upper case K). Type

```
      C ← 'ECONOMETRICS'
      C
ECONOMETRICS
```

{enclose the string of characters with a single quote at each end

Review Character Arrays

You could also store the characters of a course name and number:

```
      C ← 'ECONOMETRICS 801'
      C
ECONOMETRICS 801
```

Now suppose that you wanted to alter the string, changing just the course number. You might try

```
      C[2] ← 802
```

This would yield

DOMAIN ERROR

```
DOMAIN ERROR
      C[2] ← 802
      ∧
```

or, you might try

```
      C[2] ← '802'
```

and you would get

LENGTH ERROR

```
LENGTH ERROR
      C[2] ← '802'
      ∧
```

Let's see what has gone wrong. How many elements are contained in C?

You can ask the computer:

```
      ρC
16
```

Each letter, each number, and all blanks (in this case one) are separate elements. Remember that the ρ command tells us the "shape" of an array—the number of elements that it contains. Also recall that '801' is *not* a number, but is three characters, viz., 8, 0, 1. We are not able to perform arithmetic with characters.

When you attempted to replace the second element of the character vector C with the number 802, you were told that numbers are not in the domain of characters. The attempt to enter the three characters '802' into the one character element C[2] caused a length error.

To change the course number, we could change the whole array:

```
      C ← 'ECONOMETRICS 802'
```

or just the last three elements

```
      C[14, 15, 16] ← '802'
```

or just the last element

```
      C[16] ← '2'
```

The result in C is the same:

```
      C
ECONOMETRICS 802
```

Suppose that you had a vector of course titles

```
      A ← 'ECON MATH MONEY'
```

and a vector of course numbers, say

```
      B ← 802 801 800
```

Keep in mind that A is a character array and B is a numeric array. Now, how would you display

```
      ECON 802
```

You might try

```
      A[1 2 3 4], B[1]
```

and it would yield:

```
DOMAIN ERROR
      A[1 2 3 4], B[1]
```

Display Numeric and Character Data on One Line with ;

Unfortunately, you can't mix characters and numeric data. However, there are ways around the problem. You can *print* or *display* mixed character and numeric data by using the semicolon ; (upper case comma on an

APL keyboard):

```
A[1 2 3 4]; B[1]
ECON802
```

It might be tempting to try to form an array which could be *used* as an array. Type

```
D←A[1 2 3 4]; B[1]
```

The computer displays

```
ECON802
```

and if you did not *check* to see the result you might believe that the array D contained what you want. Try

```
D
```

The computer responds

```
ECON
```

The numeric portion is lost.

You might experiment by reversing the order and see if you obtain the same result. Thus, with mixed numeric and character data, we can use the semicolon to *print* the mixed array as output, but we cannot store it that way.

Monadic Format ⍕

It is possible to convert numerical data to character information with a command called *Monadic Format*. The symbol for the operation is ⍕; it is formed by overstriking ⊤ (upper case *N*) and little ∘, called jot (upper case *J*). In our example,

```
A[1 2 3 4], ⍕ B[1]
ECON802
```

(⍕ transformed B[1] into a character array and comma catenates the two character arrays.) You can form a new character array this way.

```
D ← A[1 2 3 4], ⍕ B[1]
D
ECON802
```

But be careful. D is a character array, not a numeric array. Try

```
D + 1
DOMAIN ERROR
D + 1
∧
```

7.2 Entering Data Inside a Function

Entering Data Inside a Function

When you read Chapter 4, you may have wondered how you can enter data into a function after the function has been defined. Until now, in executing functions, all the data have been stored in the machine *prior* to execution. If you want to enter data "inside" a function, that is, without first specifying arguments and then defining global variables to be used in the function, the following procedure does the job.

Before we write a function that uses input in this special way, let's write a function to compute the arithmetic mean. You did this in Chapter 4, but let's do it again.

```
∇ XBAR ← MEAN X
[1] XBAR ← (+/X) ÷ ρ,X ∇
```

To use the function, we store values into the array labelled *X*:

```
X ← 10 30 5 71 15.2
MEAN X
26.24
```

To do it again, we could enter new data—say,

```
B ← 3 5 4.1 .2031
X ← X, B
MEAN X
15.945
```

or

```
MEAN B
3.0758
```

Input Continuation Using ,⎕

Suppose that you had more data than could fit on a line. You could type ,⎕ (upper case *L*, called "QUAD") after one of the numbers. For example,

```
LONG←7 11 232 152,⎕
```

Then press RETURN, and the computer responds:

```
⎕:
```

Then you could type

```
5 21 31 68
```

and press RETURN. The array *LONG* is

```
LONG
7 11 232 152 5 21 31 68
```

```
MEAN LONG
65.875
```

Using ,⎕ gives you an easy way to input data into an array when the amount of data exceeds the line length of the terminal.

With reference to our mean function, notice that first we changed the data and then we executed the function. Why not enter the data inside the function? To try this, enter the function

Numeric Input
Quad-Input
A←⎕

```
∇ XBAR ← NUMEAN
[1] X ← ⎕
[2] XBAR ← (+/X) ÷ ρ,X
[3] ∇
```

To execute NUMEAN:

```
NUMEAN
```

(you type, followed by RETURN or EXECUTE)

```
⎕:
```

(the computer responds by printing quad, to indicate that it is waiting for your data.) You now type:

```
1 3 8 13 14 12.732
```

and the computer replies

```
8.622
```

Now try typing

```
X
1 3 8 13 14 12.732
```

We now see that ←⎕ enables us to define a global variable; *X* is global in the above function. NUMEAN can also be used with the previously defined global variables. Try

```
Y ← 1 2 3 4 5 6 7 8 9
NUMEAN
⎕: Y
5
```

which is the mean of *Y*.

You might be thinking, "Nice, but so what? I have to type in as much information." True, but what if we add the unconditional transfer? The program could be

```
∇ XBAR←BETRYET
```

Making a Program Interactive Using ⎕ and Branching

```
[1]     X←⎕
[2]     XBAR←(+/X)÷ρ,X
[3]     XBAR
[4]     →1
[5]     ∇
```

Now try it:

```
      BETRYET
⎕:
      1 2 4 5 12
4.8
⎕:
      2 .12 .034 34.1 71
21.4508
⎕:
```

Terminating Quad Input with →

Wait! How do we stop it? The machine will keep asking you for input. You might try to type in such words as STOP, END, etc., but you will have no luck. Try using "branch," which is the right-pointing arrow → (keyed by upper case ←). Now you are out of the function.

You can see how we have developed an *interactive* program, i.e., a program that interacts with you by prompting you. It might be easier to calculate means in this new way, especially if you had a large number of separate data sets and you were unsure about exactly how many sets there were.

Character Input Quote-Quad Input A←⍞

What about character information? We can handle it in almost the same way, but we must use "quote quad," ⍞ [type quad (upper case *L*), backspace, and then quote (upper case *K*)]. The new program might be

```
      ∇ BESTYET
[1] I ← 0
[2] AGAIN: I ← I + 1
[3] DISP ← ⍞
[4] X ← ⎕
[5] XBAR ← (+/X) ÷ ρ,X
[6] 'THE NAME OF THE DATA SET IS ' ; DISP
[7] 'NUMBER OF DATA SETS READ THUS FAR = ' ; I
[8] 'THE MEAN = '; XBAR
```

```
[9] → AGAIN
[10] ∇
```

Try it:

```
BESTYET
&&& . . . &
```

Notice that the response for ⍞ is not the same as for ⎕; here we get a "blank" response! (Some systems respond with a blinking cursor, underscore, or color.)

You name the data set by typing

```
    CHICAGO SMSA
⎕:
```

Then enter the data:

```
    33.7  27.3  31.4  84.2  33.9
    THE NAME OF THE DATA SET IS CHICAGO SMSA
    NUMBER OF DATA SETS READ THUS FAR = 1
    THE MEAN = 42.08
```

And again:

```
    SAN FRANCISCO SMSA
⎕:  34.2 41.2  48.7  84.3  44.2
    THE NAME OF THE DATA SET IS SAN FRANCISCO SMSA
    NUMBER OF DATA SETS READ THUS FAR = 2
    THE MEAN = 50.52
      AAA
⎕:
```

Terminating Input

Terminating QUAD and Quote-Quad Input Request:
→ for ← ⎕
and
O backspace U backspace T for ⍞

How does one get out of the above program? You could type any character for the name. Then type → for the numeric data. To terminate the request for quote-quad input you enter *O*, then backspace, *U*, then backspace, *T*, and then press return. This will interrupt the execution of your function. To double-check, type

```
)SI
```

(SI = state indicator: indicates which routines are "suspended," i.e., still trying to finish execution, and where the suspension occurs. A blank response by the computer means nothing is suspended.)

Clearly this is not the only, or even the best, way to handle this problem. One advantage of introducing the problem in this way was that it gave you another useful "emergency" tool in your APL tool kit. But how else might we handle this problem? Another method, which we have used before, is the conditional branch. We can instruct the user that when he is finished entering data to calculate means, he is to instruct the computer to finish the operation by typing in

```
      FINISHED
```

Now all that we have to do is to insert after statement [3] a conditional branch that instructs the computer to leave this routine when it encounters the characters *FINISHED*. Consider the following:

```
      [3.5]→EXIT×⍳8=+/'FINISHED'=8⍴DISP

      [10] EXIT:→0
```

When the character string *FINISHED* is read and stored in *DISP*, statement [3.5] tells the computer to go to the statement named EXIT. The latter statement simply terminates the function's execution. If *DISP* does not contain *FINISHED*, then `'FINISHED'=8⍴DISP` is an array of one to eight zeros. Applying +/ to this array (character by character) yields the number less than 8: the logical comparison 8 = some number less than 8 also yields 0, and ⍳0 gives an empty array. APL interprets "go to an empty array" as a statement to be ignored.

Let's proceed to alter BESTYET.

```
      ∇ BESTYET[3⎕]

[3][3.5]→EXIT×⍳8=+/'FINISHED'=8⍴DISP

[3.6][10] EXIT:→0∇
```

Don't dismay. We know that the editing of the function is not clear to you yet. You will learn exactly how to do it in the next chapter. It is presented here so that you can edit the function and understand the main points which concern entering data and branching.

Now let's retry the function.

```
      BESTYET

NEW YORK SMSA

⎕:

      1 2 3 4

THE NAME OF THESE DATA IS NEW YORK SMSA

NUMBER OF DATA SETS READ THUS FAR = 1

THE MEAN = 2.5

DETROIT SMSA

⎕:
```

```
1 2 3 4 5 6
THE NAME OF THE DATA IS DETROIT SMSA
NUMBER OF DATA SETS READ THUS FAR = 2
THE MEAN = 3.5
FINISHED
& & & &....&
```

7.3 Saving Your Workspace When Using the Computer Terminal

Saving Workspace

If you are working at a terminal, you can preserve your workspace so that your functions and variables will be available to you at your next session. Just type

)CONTINUE

```
)CONTINUE
```

and the computer will respond

```
08:21:31  01/15/80
```

indicating that your workspace was stored in your personal APL library at 8:21:31 on 15 January, 1980. You now have stored one workspace in your private library, and the *)CONTINUE* systems command also disconnected you from the system. It is just as if you issued the *)OFF* command that you used in Chapter 1, with the exception that you have preserved your workspace in a private library. When we first introduced workspaces, we said that they could be thought of as a part of the computer that was allocated to the individual APL user. Now we can expand on that and think of a workspace as a file. A computer file is an electronic analogue of a manila file folder. So an APL workspace can be retrieved from the "file cabinet," loaded, updated, saved, and erased. The name of the file is the name of the workspace.

)CONTINUE: PASSWORD

You can establish a password in the same way in which the *)OFF* command was used. For example,

```
)CONTINUE:FARM
```

However, when you log-on, you must use the *new* password:

```
)1984:FARM
```

But now the system responds

```
062*      08:16:01      01/16/80
OPR: SYSTEM AVAILABLE TO 22:30
SAVED     08:21:31      01/15/80
```

The last line would have been *CLEAR WS* if your previous signoff was)*OFF*. But with the)*CONTINUE* command, your workspace is exactly as it was when you issued the command. You might like to know that the computer automatically uses this routine if your terminal is inadvertently disconnected from the system. When you subsequently reestablish communication, your workspace is also reestablished, just as if you had issued the)*CONTINUE* command.

As a matter of convenience, you may use

)*CONTINUE HOLD*

)*CONTINUE HOLD*

In this case the APL system recognizes the word *HOLD,* and the communications line to the computer is held open for approximately 60 seconds. This saves the next user the bother of reestablishing the communication link with the computer. If you are going to use this command, it is recommended that the new user sit at the terminal and log-off the old user. This avoids a computerized version of musical chairs. However, don't just leave your terminal on in the active mode, waiting for someone to log you off. You are probably being charged for this connect time. Also, someone could log you off with a new password known to him and unknown to you.

)*WSID*
)*WSID ID*

It is possible to save a workspace without logging-off. The first step is to identify the workspace with a name. This is done with the)*WSID* (Work-Space IDentification) command.

)*WSID MONEY*

WAS CLEAR WS

Your workspace is renamed *MONEY.* The computer responded with *WAS CLEAR WS*, indicating the former name of the workspace. The next step is to actually save your active workspace. The command is

)*SAVE ID*

)*SAVE MONEY*

Your active workspace is saved in your private library under the name *MONEY.* After you issue the systems command)*SAVE MONEY*, the computer responds

9:01:21 01/16/80

indicating that the workspace was saved at 9:01:21 on 16 January, 1980. You can proceed with more computing after issuing the)*SAVE* command.

)*LOAD ID*

To load the saved workspace from your private library into your active workspace, you issue the command

)*LOAD MONEY*

The active workspace (it could be a clear ws) that existed before the command was issued is replaced by the workspace named MONEY. The computer responds with

SAVED 9:01:21 01/16/80

telling you when the workspace *MONEY* was saved. Since it is possible to save a number of workspaces, you may want to know the names of the workspaces stored in your private library. The command is

)LIB

)LIB

and the computer responds

CONTINUE

MONEY

This indicates that you have two stored workspaces in your library—one named *CONTINUE*, the other named *MONEY*.

Whenever you issue the)CONTINUE command, the workspace is stored under the name *CONTINUE*. This workspace is automatically loaded when you sign on. If you want to save your workspace and have it automatically loaded, you can execute the following command:

)SAVE CONTINUE

)SAVE CONTINUE

This command saves the active workspace under the special name *CONTINUE*. It is as if you typed)CONTINUE, but you have not logged-off the computer.

So far we have replaced the old workspace with the new one by the use of the)LOAD command. Another way to update a workspace is to add one workspace to another or to add functions or variables to an existing workspace. This is accomplished with the)COPY command. For example, you may have a number of functions stored in one workspace and a number of data sets stored in another workspace. The data (variables) and the program (functions) must be in the same active workspace to perform computations. Suppose that you have a function named *REGRE*, which was stored in workspace *STAT*. You could load it by

)LOAD STAT

Suppose that the data are stored in the workspace named *DAT* under the variables names *Y* and *X*. You can add the data from *DAT* by

)COPY

)COPY DAT Y X

The general form of the copy command is

)COPY NAME ENTITY1 ... ENTITYN

NAME is the name of the workspace, and *ENTITY* represents either a function or a variable. If the entity is omitted, all the variables and functions are added to the existing active workspace. When a function or variable conflict exists between the existing active workspace and any of those in the copy command—that is, the same variable (or function) name appears in *both* workspaces—the copy command takes precedence. The existing values are replaced by the ones in the copy command. If you want to be protected against possible unintentional conflicts or inadvertent replacements, you can protect your existing workspace by using the

)PCOPY

)PCOPY

command, which will not resolve any conflicts between the existing and copied workspaces, but will notify you of such conflicts. *)PCOPY* will only copy those entities for which no conflict exists.

After you have continued, saved, loaded, and copied a number of workspaces, it is a good idea to check what functions are in the workspace with the *)FNS* command, what variables are stored in the workspace with *)VARS*, and the name or identification of the active workspace with *)WSID*.

You can clear the workspace with the

)CLEAR

```
        )CLEAR
```

instruction. You can drop a workspace from your private library with the command

)DROP ID

```
        )DROP MONEY
```

and the computer responds with

```
        9:12:07       01/16/80
```

Private Library
)ERASE

telling you when the workspace named *MONEY* was dropped from your private library. It is now impossible to retrieve that dropped workspace. Finally, to erase a function or variable you enter *)ERASE* and the name of the item(s). You can see the result of this command by entering *)VARS* or *)FNS*.

You now know a few more ways to get data into and out of the computer. Many other ways are possible, but these basic methods will get you through many situations that you are likely to encounter in your work.

Here is an example of a terminal session that might help you to review the concepts presented in this section.

```
        CLEAR WS

        ∇MEAN
[1]     AVE←(+/X)÷⍴X                 { function to compute mean
[2]     ∇

        ∇DATAGEN                     ⎧ function generates 25 random
[1]     X←25?1000                    ⎨ numbers from 1 to 1000 without
[2]     ∇                            ⎩ replacement.

        )FNS                         ⎧ We have two functions (DATAGEN
DATAGEN MEAN                         ⎨ and MEAN) and no variables in this
        )VARS                        ⎩ active workspace.

        )WSID                        ⎧ The name of this active workspace
IS CLEAR WS                          ⎩ is CLEAR WS.

        )SAVE CASH                   ⎧ We saved the workspace under the
08:42:41 01/18/80                    ⎩ name CASH.
```

```
      )CLEAR
CLEAR WS
      )FNS
      )VARS
      )LIB
CASH
```

We clear the workspace. Computer responds that we have a cleared workspace, and we have no variables or functions in the workspace. We have one stored workspace named *CASH*.

```
      )COPY CASH DATAGEN
SAVED 08:42:41 01/18/80
```

We copy the function Datagen into the active workspace. Remember, we have another function called *MEAN* in the stored workspace *CASH*.

```
      )FNS
DATAGEN
      )VARS
```

We check this and note that only *DATAGEN* was indeed transferred to the workspace.

```
      )WSID
IS CLEAR WS
```

The name of the active workspace is not altered by the copy command.

```
      DATAGEN
```

Execute the function and list the result.

```
      X
132 756 459 533 219 48 679 680 935
384 520 831 35 54 530 672 8 67 418
687 589 931 847 527 92

      )FNS
DATAGEN
      )VARS
X
```

The effect of this has been to create a variable *X* in the workspace.

```
      )SAVE COINS
08:45:13 01/18/80
```

We save this workspace under the name *COINS*.

```
      )LIB
CASH
COINS
```

Note that the workspaces *COINS* and *CASH* are not the same.

```
      DATAGEN
```

We have executed the *DATAGEN* program again. The storing of a workspace does not change the contents of the active workspace.

```
      X
654 416 702 911 763 263 48 737 329
633 575 992 366 248 983 723 754 652
73 632 885 273 437 767 478
```

```
      )LOAD COINS
SAVED 08:45:13 01/18/80
```
We load the workspace *COINS* into the active workspace

```
      X
```
and list the values of *X*.

```
132 756 459 533 219 48 679
680 935 384 520 831 35 54
530 672 8 67 418 687 589
931 847 527 92
```
Note that the loading of *X* values from *COINS* replaced the existing values in the workspace.

```
      )SAVE CASH
NOT SAVED, THIS WS IS COINS
```
The computer will not allow us to replace the values in a stored workspace with those in an active workspace if the *WSID*s are different.

```
      )WSID
IS COINS
```
Our workspace name is *COINS* from our last load command.

```
      )WSID CASH
WAS COINS
```
We change the name of the active workspace

```
      )SAVE CASH
08:49:00 01/18/80
```
and save it (thus replacing the previous version stored in *CASH*).

```
      )DROP COINS
08:52:14 01/18/80
```
This removes *COINS* from our private library.

```
      )OFF

      LOG OFF 08:52:32  01/18/80

      END OF SESSION
```

You can log-on and see that *CASH* is still stored on your private library. If you had logged-off with the)*CONTINUE* systems command, you would have a second copy of *CASH* in your private library under the name *CONTINUE*. It would be automatically loaded when you logged-on again.

Saving Workspaces on Microcomputer

You are now at the stage where you can experiment with passwords and the)*PCOPY* command in order to see how they work in practice. Appendix D contains an explanation of one way to save your workspace on a microcomputer.

Summary

We reviewed the assignment of numeric and character data. *DOMAIN ERROR* was generated when an attempt was made to catenate the two types of data. *LENGTH ERROR* was generated when an attempt was made to assign more than one character to an element in a character array. Data can be entered into the computer from an executing function. Numeric data are entered via ⎕ and character data are entered via ⍞. When you want to enter more data than can be held on a line of your terminal, enter ,⎕ and press return.

You can save your active workspace by entering the systems command)CONTINUE. When you log-on the next time, this workspace, rather than CLEAR WS, comes up. By typing)CONTINUE:PASS you will have to use the password PASS when you log-on. Other uses of CONTINUE were also discussed.

Another systems command is)WSID; it allows you to display and change the name of your workspace.)SAVE ID allows you to save the current active workspace under the name ID. To use this workspace you enter)LOAD ID. You could bring part of this workspace to your CLEAR WS or to your active workspace by entering)COPY ID Y X. This would bring the Y and X (either functions or variables) from workspace ID to your current workspace. It is as if you loaded only X and Y. If you already had a function or variable named X or Y, the)COPY command would cause the existing item to be replaced. The)PCOPY instruction copies those items for which no conflicts exist and informs you of conflicts.

To clear your whole workspace enter)CLEAR, to erase functions or variables from a workspace enter)ERASE X Y, and to drop a workspace from your library enter)DROP ID. The remaining workspaces can be displayed by)LIB.

Exercises

APL Practice

1. Let the variables A, B, C, D be

```
A←1 2 3 4

B←5 6 7 8 9 10

C←'A B C D'

D←'E F'
```

(A) Try to predict the result before entering the following operations on the computer:

a) A,B
b) A,B,B
c) A,B[3],A[4]
d) C,D
e) C;D
f) C,C;B[4]

(B) Enter the following and predict the result. If you get an error explain why.

a) A←A,B
b) A←A,B,B
c) A←A,B[3],A[24]
d) C←C,D

e) C←C;D
f) C←C,C;B[4]
g) C←C,⍕ B[5]
h) C←C&B
i) ⍴B←B,B

2. Enter the following function:

```
    ∇ MEAN←MEAN
[1] HEADER←⍞
[2] DATA←⎕
[3] MEAN←+ /DATA÷ ⍴ DATA ∇
```

(a) Execute the function.
(b) Use any name you choose for a header.
(c) Compute the mean of: `1 2 10 ¯4 6`.
(d) Compute the mean of `(7⍟8.3)+ ⍳ 100`.

3. Display the names of all the functions in your workspace.

4. Display the names of all the variables in your workspace.

5. Display the name of your workspace.

6. (a) Why is the function MEAN computationally inefficient?
(b) How would you write it to make it more efficient?
(c) What would be the effect of putting a comma between ρ and DATA?

7. In line 10 of BESTYET what would the following produce and why?
(a) `→` (c) `→⍳0`
(b) `→0` (d) `→20`

8. (a) Save the current workspace under the name STAT.
(b) Erase all functions, variables and obtain a clear workspace.
(c) Enter a function that computes the standard deviation using the result from MEAN.
(d) Compute the mean and standard deviation of the 200 element array `(0⍳100), (23×100?100)`.

9. How would you ask the computer to:
(a) Copy the function CORR from the file whose name is STAT.
(b) Copy the variable Y from the file whose name is DATA.
(c) Copy the variable X from the file whose name is DA, given you already have another variable with the name X on your workspace.
(d) Find the correlation of X and Y, using function CORR.
(e) Store the answer in the file whose name is JE.

10. Suppose that in a file you have saved the workspace STAT which contains some functions and some variables. Write down the necessary commands that will allow you to add variable X into the file STAT.

11. In the function BESTYET we added the line

```
[3.5]→EXIT×⍳8=+/'FINISHED'=8 ⍴DISP
```

to allow for a more orderly termination of the routine. Rewrite that line using a logical function rather than +/.

Statistical Applications

1. Write a routine that will do the following:
   (a) Asks you for a number from 10 to 100. Call it N.
   (b) Take a random sample of size N from the integers 1 to 5.
   (c) Calculate the sample mean; call it $M1$.
   (d) Repeat the sampling process from (a) to find a second mean called $M2$.
   (e) Find the mean and the variance of $M1$ and $M2$ [in this case $(M1 + M2)/2$], call it $MM2$. Find the sample variance of $M1$ and $M2$ [in this case $((M1 - MM2)^2 + (M2 - MM2)^2)/(2 - 1)$], and call this $SS2$.
   (f) Repeat the sampling process from (a) to find a new mean called $M3$, and recompute the "average" mean and variance, say $MM3$ and $SS3$.
   (g) Repeat the sampling process up to the point where the difference of the sample variances of the $MM_i - MM_{i-1}$ is less than .001. How many samples does it take?

2. Let the variables X and Y be defined as `X←Y×Y←÷(¯4 ¯3 ¯2 2 3 4)` and the variables `W←Y-X`. Write a function to calculate the correlation between any two variables and use this function to verify that:
   (a) X and Y are not correlated (i.e., their sample correlation coefficient is close to zero).
   (b) W and X are perfectly negatively correlated (i.e., their sample correlation coefficient is very close to ¯1).
   (c) In general, how would you generate two discrete random variables that are perfectly negatively correlated?

3. The joint discrete distribution of two arrays A and B is:

		A						
		¯3	¯2	¯1	0	1	2	3
B	10	.05	.10	.18	.19	.04	.02	.01
	20	.01	.12	.13	.07	.06	.01	.01

(a) Compute the marginal distributions of A and B.

(b) Compute the conditional distribution of A given $B = 10$.

(c) Are A and B independent?

(d) Are A and B uncorrelated?

4. One form of the central limit theorem states that if X_i, $i = 1, \ldots, N$ are independent random variables with mean μ and variance σ^2, and $Z_i = (\bar{X} - \mu)\sqrt{n}/\sigma$ is the standardized X_i, then Z_i is distributed as standard normal (i.e., Z_i has mean value 0 and variance 1). Write a function that will give you a random sample $X_1, X_2, \cdots X_n$ of size 30 of a random variable which takes on values $1, 2, \cdots 10$, each with probability .1, and calculate $z_1 = ((\bar{X} - 5.5)/\sqrt{8.25})\sqrt{30}$. Repeat the process nine times and calculate the sample mean and variance of $Z_1, Z_2 \cdots Z_{10}$.

***5.** Write a function that will utilize the functions I NORMC X (page 85) and the function NORM D (page 84) to calculate the tables of the standard normal distribution for the values of −4 to 4 in intervals of .1. Compare your table to the standard normal tables.

6. An economist who believes in cardinal utility had to decide on one of two dates. Date A would yield 40,000 utils* if successful and 4,000 utils if unsuccessful. Date B would yield 70,000 utils if successful and 10 utils if unsuccessful. Chances of a successful date were subjectively estimated to be equal for A and B. The economist was in a quandary about what to do. (She was indifferent between the two choices.) So she decided to compute the probabilities of success that would make A the preferred choice, B the preferred choice, and A and B equally attractive on the basis of expected utility. Next, suppose that she could be assured of receiving 8,000 utils by remaining home and learning more about APL. Which of the three alternatives would she now choose?

Finally, assume that the probabilities were such that the expected utilities were in fact equal. However, the economist could spend time, and thus utils, in improving the chances of success in case B. In this case the probability of success in A might remain the same as before, and the chance of success in B would be larger than that in A. State a decision rule for spending utils to improve the probability of success. If she could buy an increase in the probability of success to 20% and reduce the probability of failure to 80% for date B at the cost of 6,000 utils, should she do it?

* A util is an arbitrary measure of satisfaction.

8

More on Functions

Thus far we have been very careful with the definitions of our functions and have cautioned you to be very careful in typing in the suggested APL expressions. Now is the time to become more adventurous in our writing of functions and to no longer worry unduly about making mistakes. One of the delights of APL for those of us who are used to programming in FORTRAN and similar languages is that correcting mistakes and errors is so much easier in APL than it is in the other computer languages. We all make mistakes, so it is reassuring to know that correcting our errors will not be difficult.

8.1 Function Display, Correction, and Editing

You will find that you will want to be able to examine your function from time to time and, if it is a long one with many statements in it, you will often want to look at only a part of the function, not at all of it. Next, you will discover that you will need to alter, if not to correct, one or more lines in your function. So we need easy ways to display a function and to alter, add, or delete lines in it.

The next most important task is to be able to figure out how and why your function does not work or does not produce the results it should. This brings us to the intriguing world of diagnostics. Let's begin with the simpler task of correcting errors we know about, before we learn how to discover the ones we don't know about.

If your typing is like ours, you made a few typing errors when entering the function statements. As long as you caught them on the line on which they were made, prior to pressing the RETURN or EXECUTE key, they were easy to correct. We discussed a method of correcting such errors on page 25. However, if you discovered the typo after the function definition was closed off, that is, after the final ∇, you had troubles. On page 61 we showed one way to correct the errors. Here are some more ways to modify your function. Let's start with your old friend from the previous chapter, *BESTYET.*

Function Display ⎕

To display your function, type

```
    ∇BESTYET[⎕] ∇                                    ← you type
[1] I←0
[2] AGAIN:I←I+1
[3] DISP←⍞
[4] →EXIT×⍳8=+/'FINISHED'=8⍴DISP
[5] X←⎕
[6] XBAR←(+/X)÷⍴X                                    computer responds
[7] 'THE NAME OF THESE DATA IS ',DISP
[8] 'NUMBER OF DATA SETS READ THUS FAR = ';I
[9] 'THE MEAN = ';XBAR
[10] →AGAIN
[11] EXIT:→0
    ∇
```

Display One or Several Lines of a Function [# ⎕] *or* [⎕ #]

As you may remember, this is how to obtain a display or listing of the *whole* function. Should you want to display only one statement in the function, for example, number three, then you can type

```
    ∇BESTYET[3⎕] ∇
```

and the computer prints

```
[3]  DISP←⍞
```

What you have done is to open the function by typing ∇*BESTYET*. Statement [3] is displayed by using the ⎕ (quad) symbol, and the function is then closed with the final ∇ (del). If you had typed

```
    ∇ BESTYET[⎕3] ∇
```

you would have received all the function statements from [3] to the end of *BESTYET.*

Suppose you want to *modify* a single statement, say [7]. You can type

```
    ∇ BESTYET[7] 'THE NAME OF DATA SET IS ';DISP ∇
```

This will replace the former statement number 7 with this new one. Another way to modify a statement is

```
    ∇ BESTYET[8⎕]                                ← you type
[8] 'NUMBER OF DATA SETS READ THUS FAR = ';I    ← computer responds
```

This is the first time that we have not closed the function. You have opened the function, the computer has displayed statement [8], and you are now able to modify it by typing

```
[8] I;'DATA SETS HAVE BEEN READ'                ← you type
[9]                                             ← computer responds
```

Remember to Close the Function

Now you must close the function by typing on line [9] the symbol ∇.

Yet another way to modify your function is to display the function and omit the closing del.

```
      ∇ BESTYET[⎕]
```

This will list your function and allow you to add a statement at the end of the function, or to modify other function statements or to insert new statements in your program. Let's start by adding a statement at the end. After listing the entire function, including the final ∇, the computer will type

```
[12]
```

and you can add:

Insert a New Line

```
[12]  'THIS STATEMENT CANNOT BE REACHED'
```

Next we could modify statement [9] by

```
[13]  [9⎕]                              ← you type to display line 9
[9]   'THE MEAN = ';XBAR                ← the computer responds
[9]   'THE ARITHMETIC MEAN = ';XBAR     ← you type a correction
[10]                                    ← the computer responds with
                                          the next line number
```

To insert a statement between statement [2] and statement [3] you would type

```
[10]  [2.2]'ENTER THE NAME OF THE DATA SET'
```

and you could add

```
[2.3] 'THIS STATEMENT HAS NO PURPOSE'
[2.4]   ∇
```

Before we delete the last statement we just inserted, we should list the function:

```
          ∇BESTYET[⎕]
      ∇BESTYET
[1]  I←0
[2]  AGAIN:I←I+1
```

```
[3]  'ENTER THE NAME OF THE DATA SET'
[4]  'THIS STATEMENT HAS NO PURPOSE'
[5]  DISP←⍞
[6]  →EXIT×⍳8=+/'FINISHED'=8 ⍴ DISP
[7]  X←⎕
[8]  XBAR← (+/X)÷⍴X
[9]  'THE NAME OF THE DATA SET IS';DISP
[10] I;'DATA SETS HAVE BEEN READ'
[11] 'THE ARITHMETIC MEAN = ';XBAR
[12] →AGAIN
[13] EXIT:→0
[14] 'THIS STATEMENT CAN NOT BE REACHED'
       ∇
[15]
```

Statements Automatically Renumbered

(Notice that the statements we entered as [2.2], [2.3] are renumbered as [3], [4], and all subsequent line numbers are increased by two: another reason for using line *labels* in branch statements!)

The last (useless) statement we added is now Number 4; we added it in order to show you how to *delete* a line. On some systems we can type

Deleting a Line in a Function

```
[15]  [Δ4]
```

On other systems this may not work. Instead of [Δ4], type [4] and hit LINEFEED followed by EXECUTE or, on some terminals, ATTN followed by EXECUTE.

If we decided that we want to prompt the user of this function when data are entered, we might insert

```
       ∇ BESTYET [14⎕]
[14]  [4.1] 'ENTER DATA, SEPARATE OBSERVATIONS BY BLANKS AND
             PRESS EXECUTE AFTER LAST SAMPLE'
[4.2] ∇
```

Display the function as rewritten once again, and note the renumbering of the statements. Renumbering would cause problems if your branch statements referred to a specific line number, so that is why it is best to label statements that receive branches.

For example, consider this very simple function and its change. Enter

```
       ∇ TRIAL
```

```
[1]   →2
[2]   'END' ∇
```

← first statement instructs computer to execute 'END' by branching to [2]

Now alter the function as follows:

```
      ∇ TRIAL[3⎕]
[3] [1.1]A←1
[1.2] ∇
         ∇ TRIAL[⎕]  ∇
      ∇ TRIAL
[1] →2
[2] A←1
[3] 'END'
      ∇
```

After change, lines are renumbered but the *branch statement is not,* so the instruction in [1] is no longer correct.

Now try using the function with the data as before. You can treat the function header as if it were another statement with line count 0 (remember: zero—not alphabetic "oh").

```
      ∇BESTYET[0⎕]
[0]  BESTYET
[0]  BSTYET
[1]   ∇
```

← you type

← computer responds—note the next number is *not* put in until you hit RETURN

← you type

We have renamed the function. We could also change the "type" of the function by altering the header in various ways. For example,

```
      ∇BSTYET[0⎕]
[0]  BSTYET
[1][0] XBAR←BSTYET
[1]  ∇
```

To erase the whole function, type)*ERASE BSTYET*. To reassure yourself that it is gone, type

```
      BSTYET
      VALUE ERROR
      BSTYET
      ^
```

or type)*FNS* and check to see that BSTYET is no longer listed.

After you have gained a little practice you will find that your ability to edit your functions will become second nature. Of course, it is all very well to be expert at editing your functions and correcting errors if you know what has to be done. The real challenge is finding out what went wrong with your function; even more challenging are those cases wherein you are not sure anything is wrong, but it might be! We now come to computer detective work: how to find errors and mistakes. As in all detective work, there are some rules to know and some recommended procedures to follow, but after that you are on your own in learning the arcane art of computer error detection. Let's begin with some useful procedures.

8.2 Diagnostic Procedures

Diagnostics

Define the function

```
      ∇ NEW←ECON X
[1]   NEW←ρX÷(+/(1÷X))
[2]   GEO←(×/X)*(1÷ρX)
[3]   ∇
```

What are the expected results? When the vector *X* containing 2 4 8 is executed, the harmonic mean should be 3.4286, and the geometric mean should be 4.

What does the computer produce?

```
      X←2 4 8
      ECON X
      3
```

Is this satisfactory? Of course not, but what about the variable *GEO?*

```
      GEO
      4
```

This is OK, but is this "test" adequate? The latter question is not easily answered in general. First, let's alter *ECON* to print *GEO* every time it is executed.

```
        ∇ ECON[⎕]
      ∇ NEW←ECON X
[1]   NEW←ρX÷(+/(1÷X))
[2]   GEO←(×/X)*(1÷ρX)
      ∇
[3]   ⎕←GEO ∇
```

```
      ECON X
4
3
```

Is this the order you expected? If not, remember that when the function is executed, *GEO* will be printed at line [3] and then *NEW* will be printed upon completion of execution.

Let us now work through our routine one step at a time in order to figure out what we have done wrong.

1. Let's write the mathematical expression we want.

$$NEW = \frac{N}{\frac{1}{x_1} + \frac{1}{x_2} + \cdots + \frac{1}{x_n}}$$

for N observations on vector X. Compute the answer by hand:

$$NEW = \frac{3}{\frac{1}{2} + \frac{1}{4} + \frac{1}{8}} = 3.4286$$

2. Are the data correct?

```
      X
2 4 8
```

Yes

3. Now that we are reasonably sure that the error is in line [1], let's find out exactly where it is. Try doing each part of the line separately:

a. (1÷X)

```
0.5 0.25 0.125
```

b. (+/(1÷X))

```
0.875
```

c. ρX

```
3
```

d. ρX÷(+/(1÷X))

```
3
```

Now we have found a problem. Two components—each apparently correct—fail to produce the correct result when combined. Remember that the computer operates from *right to left*. So let's reexamine the entire line. If you do, you will recognize that you need parentheses around ρ*X*. Consider

```
      X÷(+/(1÷X))
2.2857 4.5714 9.1429
```

```
ρX÷(+/(1÷X))

3
```

The function is producing a result which we would obtain if we had written

```
ρ (X÷(+/(1÷X)))
```

So all that we did in line [1] was to find the number of elements in the string $(X \div (+/(1 \div X)))$. What we really want is

```
(ρX)÷(+/(1÷X))

3.4286
```

The Trace Function

Trace; T∆

Now let us suppose that our function is a little more complicated and that we do not know where our problem initially occurs. In order to locate the source of the error, we can use the "trace" function. For example, suppose we want to print out the intermediate results in a routine called *DSTAT*. In particular let us suppose that we want to examine the results of the computations performed at lines [1], [2], and [4]. We can do this by typing

```
T∆DSTAT←1 2 4
```

where ∆ is upper case *H*. This allows us to see the results of each statement listed. The numbers on the right of the ← are the statements we want to trace. In order to implement the trace operation, having first specified what is to be traced, we merely execute the function. Before doing that, let us display the function *DSTAT*, which we defined in Chapter 5 and which you might have stored in a file after reading Chapter 7.

```
      ∇ DSTAT[⎕] ∇

    ∇ DSTAT X

[1]  R←(MAX←X[ρX])-MIN←(X←X[⍋X])[1]

[2]  SD←(VAR←(+/(X-MEAN←(+/X)÷N)*2)÷(N←ρX)-1)*0.5

[3]  MD←(+/|X-MEAN)÷N

[4]  MED←0.5×+/X[(⌈N÷2),1+⌊N÷2]

[5]  'SAMPLE SIZE'

[6]  N

[7]  'MAXIMUM'

[8]  MAX

[9]  'MINIMUM'
```

```
[10] MIN
[11] 'RANGE'
[12] R
[13] 'MEAN'
[14] MEAN
[15] 'VARIANCE'
[16] VAR
[17] 'STANDARD DEVIATION'
[18] SD
[19] 'MEAN DEVIATION'
[20] MD
[21] 'MEDIAN'
[22] MED
     ∇
```

We need an array of numbers on which to operate, so let us define

```
     W←9 10 2 8 12 0 1 5
     DSTAT W
DSTAT[1] 12
DSTAT[2] 4.517821852        } called by the Trace function
DSTAT[4] 6.5
SAMPLE SIZE                 ←begins the programmed output
8
MAXIMUM
12
MINIMUM
0
RANGE
12
MEAN
5.875
VARIANCE
```

```
20.41071429

STANDARD DEVIATION

4.517821852

MEAN DEVIATION

3.875

MEDIAN

6.5
```

What the trace function does is to print out certain intermediate values in a routine, as specified by the array in the statement *T∆FUNCTION NAME←ARRAY*. What is printed on each line is the final (leftmost) variable defined by the operator ←. That is why, in tracing line [1], we get *R* (the range), but neither *MAX* nor *MIN*.

The trace function will remain in force until *you remove it*. This is done by typing in

```
    T∆DSTAT←⍳0
```

That is, we instruct the computer to trace the *DSTAT* line numbers contained in an empty array.

Before leaving trace, try:

```
    T∆DSTAT←0
```

The Stop Operator

Another very useful diagnostic tool is the stop operator, which is used in a manner similar to the trace operator. The stop operator is activated by typing

Stop; S∆

```
    S∆DSTAT←1  4
```

and used by calling

```
    DSTAT W
DSTAT[1]
```

What happens is that when *DSTAT* is executed, the routine stops automatically at the line *before* the one indicated. In our example, the function *DSTAT* is *ready* to execute the first line, but has not yet done so. When a routine is suspended in this way, you can do a host of other calculations, print results, etc. The function remains suspended while these other operations are being performed.

Try typing in

```
    R
```

```
VALUE ERROR
      R
      ∧
      MAX
VALUE ERROR
      MAX
      ∧
      MIN
VALUE ERROR
      MIN
      ∧
      X
9 10 2 8 12 0 1 5
```

These results show that because the routine has been stopped before *R*, *MAX*, and *MIN* have been defined, we get value errors if we try to display them. The output also shows we are inside the routine, since the dummy variable *X* contains the array *W*.

In order to execute line [1], we type

```
      →1
      DSTAT[4]
```

and not only line [1], but also lines 2 and 3 will be executed. However, line [4] will *not* be executed, since we have a stop at [4]. If we had typed →2 instead of →1 above, execution of line [1] would not have been carried out, and so the variables *R*, *MAX*, and *MIN* would still not be defined. The instruction →2 instructs the computer to go to line [2], execute it, and then continue, executing line by line, to the next stopped line.

We happen to know that at the moment our routine *DSTAT* is suspended just before line [4], but after a confusing terminal session, we may not be so sure of ourselves. Fortunately, in APL we have an easy way to find out. Type in the system command

)SI
Suspended Functions

```
      )SI
DSTAT[4]*
```

SI stands for state indicator; it tells us which routines are currently suspended, as well as where the suspensions have occurred. As you see from our example, the position is given by the next line to be executed. That is, line [4] of *DSTAT* is the next statement to be executed.

Pendent Functions

You may wonder: Why the asterisk, *? The reason is that it distinguishes between *suspended* functions and *pendent* functions. Any APL expression can include another program function. When an APL statement cannot be completed because a program function it called is suspended, the former function is called pendent and the latter is called suspended. If statement [2] in function *A* used function *B*, and statement [3] in function *B* used function *DSTAT*, which contained an undefined value at statement number [1], then the state indicator would be

```
)SI

DSTAT[1]*

B[3]

A[2]
```

Suspended functions have a * and pendent ones don't.

Both suspended and pendent functions can cause you a lot of mysterious errors if you are not careful, so the best advice is to get rid of them as soon as possible. Before we show you how to do that, we will point out that functions usually get suspended not by the use of stop commands, but because an error is encountered which halts execution—for example, a length, or value, or domain error.

Consequently, if your routine is stopped in its execution by an error and you decide to reexecute after putting in a stop or a trace command, remember *first* to get rid of the suspended and pendent functions.

This is done very simply. Type in

```
→
```

and repeat it as many times as is necessary to get a *blank* response to your query

```
)SI
```

System Command)SIV

)SIV

Another useful system command is)SIV. This command provides the same information as)SI, but in addition it gives you a list of the dummy variables that appear in the header of the suspended function.

8.3 A Case Study in Program Development and the Location and Correction of Program Errors

Debugging

In order to give you some idea of the problems involved in finding and removing errors from a program, a process which is known as "debugging," we are going to work through a case study of writing an APL routine for a "live" practical problem.

As you will soon see, the way in which this program was written is a classic example in many respects of how *not* to write a routine; however, it is a simple example, and at the moment that's what counts most.

This example concerns the effectiveness of a new drug for the treatment of duodenal ulcers. As part of the study the question was asked, "Does this drug have an effect on the work hours lost due to the disease?" Medical researchers collected data and decided to test the hypothesis using the Student t distribution. Since the researchers had access to APL, they decided to write an APL program to compute the t statistic. Here is how they did it.

First, they went to a textbook which contained the formula for the statistical test they wanted, as well as a test problem with a computed result.* The equation for the test statistic they used was:

Student's t

$$t = \frac{\bar{X}_1 - \bar{X}_2}{\sqrt{\dfrac{(N_1 - 1)s_1^2 + (N_2 - 1)s_2^2}{N_1 + N_2 - 2}} \sqrt{\dfrac{1}{N_1} + \dfrac{1}{N_2}}}$$

where $\bar{X}_1$, $\bar{X}_2$, s_1^2, s_2^2, N_1, N_2, are the respective sample means, variances, and sample sizes.

In the textbook problem, two small random samples were drawn—one from freshmen, the other from seniors—of the amounts of money that these students had on their persons. The null hypothesis was that the average amount of money held by freshmen was equal to that held by seniors. The alternative hypothesis was that the two means were unequal. The following statistics were cited:

Variable	Symbol	Freshmen	Symbol	Seniors
Average Amount of Money	X_1	\$1.28	X_2	\$2.02
Standard Deviation	S_1	0.51	S_2	0.43
Number of Observations	N_1	10	N_2	12

The critical value of the t statistic at the 5% significance level for $(N_1 + N_2 - 2) = 20$ degrees of freedom is ± 2.086 for a two-tailed test. If the computed t value is larger than the positive or smaller than the negative value, we would reject the null hypothesis at this level.

Here is a listing of the program written by the medical researchers:

```
      ∇ TTEST[⎕] ∇

     ∇ T←TTEST
[1]   SS←S1*2
[2]   SSS←S2*2
[3]   NO←N1-1
[4]   NT←N2-1
[5]   A←((NO×SS)+(NT×SSS))÷N1+N2-2
[6]   A←A*0.5
[7]   B←((1÷N1)+(1÷N2))*0.5
```

* The text was Hamburg, Morris, *Statistical Analysis for Decision Making* (New York: Harcourt, Brace and World, Inc., 1970), pp. 347–348.

```
[8]  C←A×B
[9]  T←(X1-X2)÷C
      ∇
```

The first two statements convert the standard deviations provided in the examples to the variances. The next two statements each subtract one from the number of observations. Statement [5] computes the ratio in the first radical; number [6] takes its square root. Statement [7] computes the value of the second radical and number [8] computes the denominator of the ratio. The last statement computes the t statistic itself.

Notice that they took each step and broke it down into small parts, each part getting one line. This procedure creates more lines, but it makes each line very easy to understand. They could have put many of these steps on one line. However, if one does that, then locating possible errors can become more difficult.

The next step was to store the *test* data in their active workspace:

```
      X1←1.28
      X2←2.02
      S1←0.51
      N1←10
      S2←0.43
      N2←12
```

and execute the function:

```
      TTEST
¯3.6953
```

Since the result was the same as that reported in the text, the medical researchers concluded that the program was functioning correctly. The next step should have been to compute other test values and see if they worked, too, but the researchers didn't do this.

How Mistakes Begin

Next the researchers decided that they would have to compute the standard deviation themselves from their raw data. To accomplish this, they constructed a second function, although the routine could have been incorporated in the *TTEST* program. The textbook formula for the standard deviation was

$$STDEV = \sqrt{\frac{\Sigma_1^N (X_i - \bar{X})^2}{N - 1}}$$

and their APL program to do this is listed below.

```
          ∇ STDEV[⎕] ∇
        ∇ S1←STDEV X
[1]    XBAR←(+/X)÷ρX
```

```
[2]    SA← +/(XBAR∘.-X)*2
[3]    S1←(SA*0.5)÷(ρX)-1
         ∇
```

The next step was to revise the *TTEST* program in order to incorporate this calculation. Here is that revised program:

```
            ∇ TTEST[⎕] ∇
         ∇ TT←TTEST
[1]    STDEV X
[2]    N0←(ρ X)-1
[3]    X1←XBAR
[4]    SS←S1*2
[5]    STDEV Y
[6]    NT←(ρ Y)-1
[7]    X2←XBAR
[8]    SSS←S1*2
[9]    A←((N0×SS)+(NT×SSS))÷N1+N2-2
[10]   A←A*0.5
[11]   B←((1÷N1)+(1÷N2))*0.5
[12]   C←A×B
[13]   TT←(X1-X2)÷C
         ∇
```

Let's go through this. Line [1] executes the standard deviation function. Line [2] subtracts 1 from the first sample size, line [3] renames the first group mean (so that its value will not be lost in line [5]), and line [4] computes the variance from the standard deviation computed in *STDEV*. Lines [5] through [8] repeat the process for the second group of data. The remaining part of this function is the same as before. They entered a test set of data and executed the program:

```
      Y←10 20 30 40 50
      X←1 2 3 4 5 6 7 8 9 10
      TTEST
1.0092
7.9057
¯12.981
```

The standard deviation and the t value are printed. The values obtained seemed to be in the ball park, so they proceeded to the real data.

Even at this point in our discussion you can see that the medical researchers have gotten themselves into some peculiar programming, because it would appear that they did not think ahead and plan out what they wanted to do. For example, they wrote a separate routine to calculate standard deviations, which they promptly resquared in *TTEST.* Also, if they had considered more carefully which variables should be local and which global, they would have been able to avoid having to store *XBAR* into $X1$ and $X2$ and then taking the difference.

OUTER PRODUCT (∘.g)

There is one operation in this routine with which you are not yet familiar; that is the outer product operation (∘.g) which appears in line [2] of *STDEV*. The symbol g represents any dyadic operator, and outer product is always used in the dyadic mode. The operation `(XBAR∘.-X)*2` produces an array, each element of which is $(X_i - XBAR)^2$. The same result could be obtained by `+/(XBAR-X)*2`. The latter approach to programming might be preferable to the former in that it is both easier to understand and on some systems is quicker to execute. The outer product operation (∘.−) is a most useful tool, but is probably "overkill" for our simple needs in this routine.

The data that the researchers used are:

```
      TREATMENT
1 3 4 5 5 5 5 2 3 4 1 5 0 6 2 3 5 5 5 5 2 5 5 3 5 4
      5 5 1 1 3 5 4 5 1 5 5 5 5 0
      ρ TREATMENT
40
      CONTROL
2.5 5 0 2 2.5 1 5 5 3 5 5 5 3 3 5 5 4 0 4 0 3 0 5 1 3
      1 3 5 2 5 0 1 2
      ρ CONTROL
33
```

indicating the number of work days lost for each patient in the two groups before the medication was administered. The researchers had to store these data under the names X and Y, since the programs as written can use only arrays X and Y as input (another reason for thinking ahead and considering carefully the use of dummy variables in the header).

```
      X←TREATMENT
      Y←CONTROL
```

Upon executing the program, they obtained

```
      TTEST
```

```
0.27425

0.32809

9.1397
```

and printed the means:

```
      X1
3.7
      X2
2.9091
```

Checking the Results

The means were correct (this was known from previous work), but the value for t of 9.1397 seemed to be inordinately high. Now they became suspicious. Were the data wrong? Was this a correct, even though surprising, result? Or was there something wrong with this patchwork programming?

The t statistic function looks good in the sense that it produced results that were corroborated from an outside source, but they did alter it *after* the test. However, the standard deviation program was not checked in the same way. The program was so simple that they did not run any test data. The mistake they made is an example of a very common type of error. Let us check the routine out with a simple data set that we can easily compute by using the calculator or immediate execution mode of APL.

Using APL in the CALCULATOR Mode

```
      X←ι10
      X
1 2 3 4 5 6 7 8 9 10
      STDEV X
1.0092
      XBAR
5.5
```

The mean is correct, but the standard deviation is wrong! The hand calculation is Standard Deviation =

```
      +/(XBAR-X)*2
82.5
      (82.5÷9.0)*.5
3.0277
```

We now know that there is a problem, but not where that problem occurs. So let's use our trace function.

```
      T∆STDEV←1 2 3
      STDEV X
```

```
STDEV[1] 5.5

STDEV[2] 82.5

STDEV[3] 1.0092

1.0092
```

Line [1] we have already checked by hand, and we also find that line [2] is correct. So the error must occur in line [3]. Line [3] is

```
        [3] S1←( SA*0.5)÷(ρ X)-1
```

Rethinking through this line from right to left, we discover that they computed $(\sqrt{\Sigma(X_i - X)^2})/(N - 1)$ rather than $(\sqrt{\Sigma(X_i - X)^2/(N - 1)})$. We can correct this error by editing *STDEV:*

```
        ∇ STDEV[3⎕]

[3]   S1← ( SA*0.5)÷(ρ X)-1

[3]   S1← ( SA÷((ρ X)-1))*0.5

[4]   ∇
```

Upon executing the revised function, we obtain

```
        STDEV X

STDEV[1] 5.5

STDEV[2] 82.5

STDEV[3] 3.0277

3.0277
```

which yields the correct answer.

Remember that the trace function will stay in effect until we retract it with the following command:

```
        T∆STDEV←⍳0
```

Our next step is to retry the *TTEST* program with the new *STDEV* program. We generate a second data set *Y* and run the program.

```
        X

1 2 3 4 5 6 7 8 9 10

        Y←10 20 30 40 50

        TTEST

3.0277

15.811

¯380.03
```

Our hand calculation for the standard deviation of *Y* is correct, but the value for t of -139.16 is wrong. Our hand calculation is

$$t = \frac{5.5 - 30}{\sqrt{83.269}\,\sqrt{0.3}} = -4.9019$$

Consequently, *TTEST,* which was *thought* to be correct, is in fact in error. Let us compare a trace on *TTEST* run before correcting *STDEV* with one run afterwards, and see if we get some clues as to where the error occurs.

```
      TTEST
STDEV[1] 5.5
STDEV[2] 82.5
STDEV[3] 1.0092
TTEST[1] 1.0092
TTEST[2] 9
TTEST[3] 5.5
TTEST[4] 250
STDEV[1] 30
STDEV[2] 1000
STDEV[3] 7.9057
TTEST[5] 7.9057
TTEST[6] 4
TTEST[7] 30
TTEST[8] 250
TTEST[9] 9.5588
TTEST[10] 3.0917
TTEST[11] 0.180889
TTEST[12] 0.33667
TTEST[13] ¯72.771
¯72.771
```

Next, we run the trace function on our program with the corrections:

```
      TTEST
STDEV[1]  5.5
STDEV[2]  82.5
STDEV[3]  3.0277
```

```
TTESTC[1] 3.0277

TTEST[2] 9

TTEST[3] 5.5

TTEST[4] 250

STDEV[1] 30

STDEV[2] 1000

STDEV[3] 15.811

TTEST[5] 15.811

TTEST[6] 4

TTEST[7] 30

TTEST[8] 250

TTEST[9] 9.5588

TTEST[10] 3.0917

TTEST[11] 0.10889

TTEST[12] 0.33667

TTEST[13] ¯72.771

¯72.771
```

Compare line [3] in *STDEV* with line [13] in *TTEST*. We have different inputs (from *STDEV*) but the same output from *TTEST*. If we look at statement [9] in the display of the function, we see that the results of lines [10], [11], [12], and [13] are the same, even though we have different values for the standard deviation. First we notice that *N*1 and *N*2 are not defined in the program itself. Whatever values variables *N*1 and *N*2 happen to have will be used in the program. We need statements like *N*1 ← ρ *X* and *N*2 ← ρ *Y*. But is this all? No. We also notice that the dummy variable in the header of the function *STDEV* is used as if it were a global variable! We should have received an error message when we tried to execute *TTEST*. The reason we did not is that a global variable *S*1 was defined earlier and is still floating around waiting to trap the unwary.

Problems from Careless Use of Global Variable

Here we have two more reasons why great care is needed in deciding which variables should be dummies in the header, which local, and which global. It is also a warning to keep track of the globally defined variables and to erase those no longer needed.

The rewritten program is listed below. Note in particular lines [1], [3], [5], and [7].

A Student's t Test Routine

```
     ∇ TTEST[⎕] ∇

   ∇TT←TTEST

[1]  N1← ρX
```

```
[2]  N0← (ρ X)-1
[3]  SS← ( STDEV X)*2
[4]  X1←XBAR
[5]  N2← ρ Y
[6]  NT← (ρ Y)-1
[7]  SSS← (STDEV Y)*2
[8]  X2←XBAR
[9]  A← (( N0×SS)+ ( NT×SSS))÷N1+N2-2
[10] A←A*0.5
[11] B ← ((1÷N1) +( 1÷N2))÷0.5
[12] C←A×B
[13] TT← ( X1-X2)÷C
      ∇
```

Trace Routine

Running the routine with the trace operator and the test data yields

```
     TTEST
TTEST[1]  10
TTEST[2]  9
STDEV[1]  5.5
STDEV[2]  82.5
STDEV[3]  3.0277
TTEST[3]  9.1667
TTEST[4]  5.5
TTEST[5]  5
TTEST[6]  4
STDEV[1]  30
STDEV[2]  1000
STDEV[3]  15.811
TTEST[7]  250
TTEST[8]  30
TTEST[9]  83.269
TTEST[10] 9.1252
```

```
TTEST[11] 0.54772

TTEST[12] 4.9981

TTEST[13] ¯4.9019

¯4.9019
```

These results check with hand computation at every stage, so we can now have more confidence in the correctness of this routine.

The new *TTEST* routine gives an answer of t = 1.8908 rather than t = 9.1397 in the comparison of the treatment and control data arrays that were defined on page 131.

Summary

Display of functions and use of quad, ⎕, (uppershift *L*):

∇ function name [⎕]∇ : displays the entire function
∇ function name [3⎕]: displays line [3]
∇ function name [0⎕]∇: displays header of function
∇ function name [⎕3]: displays function *from* line [3] to end
∇ function name [4] (new APL expression): changes line [4] from what it was to that specified
∇ function name [⎕]: displays function and then enters next line number ready for an additional statement (previous last line plus one)

A line may be inserted between two existing lines, say 2 and 3, in a function by using:

[13][2.2] (APL expression to be inserted)
Deleting a line, say number 4:
[13] [4] ← you enter [4] at line [13]
Hit Line Feed or ATTN key
Hit Execute key

If you use branches in your APL statements, either conditional or unconditional, use line *labels*, not numbers.
Examples:

Unconditional Branch	Conditional Branch
Use: →*EXIT*	Use: →*EXIT*×⍳*A* > *B*
Not: →10	Not: →10×⍳*A* > *B*

Computer Diagnostic Tools

Computer Diagnostic Tools:

Trace: *T*∆ function name ← l_1, l_2, . . . , l_n sets up a "trace operation" that will be activated every time the function is used. $l_1 < l_2 < l_3 \ldots$

$< l_n$ are line numbers. Upon function execution, trace will print each line number, $l_1, l_2, \ldots, l_n$ being traced, followed by the final (i.e., left-most) computed result of the line, which is usually the last variable defined (assigned a value) in each line.

Trace is removed by:

```
TΔ function name ←⍳0
```

Stop: S Δ function name $\leftarrow l_1, l_2, \ldots, l_k$ sets up a "stop operation" that will be activated every time the function is used. $l_1 < l_2 < \ldots < l_k$ are line numbers. Upon function execution, stop will print the first line number listed, l_1, *before* executing it; all prior lines will have been executed. The computer will not execute line l_1. Line l_1 or any subsequent line can be executed by typing right arrow and line number l. For example:

```
→5
```

While the function is stopped by the STOP command, arithmetic operations can be used and functions called.

Stop is removed by:

```
SΔ function name ←⍳0
```

Any function that cannot continue executing its statements, either because of a programming error or because of a STOP command, is said to be *suspended*.

Any function that cannot be executed because it calls a suspended function is said to be *pendent*.

Example of the above and the use of)SI (state indicator system command): Suppose that at line [3] function A uses function B, which is suspended at line [5].

```
)SI          ←you type
B[5]*        ←function B is suspended at 5
A[3]         ←function A is pendent at 3
```

To clear both suspended and pendent functions, type

```
→
```

as many times as necessary to get a blank response from the command)SI.

Exercises

APL Practice

1. Let's investigate some of the uses of the functions you learned in this chapter.

   (a) `5+⎕←4+3`

(b) Type `3×⎕` and execute. The computer responds ⎕:. Type 6. What do you get?

(c) Type `X←⍞` and execute. Now type 1. What is in X? Type $X = 1$ and $X =$ '1'.

(d) `X,⍞` and execute.

(e) Examine the difference between 10 ρ 'W', 'TAC' and `10ρ'W';'TAC'`.

(f) Let `T←'20'` and type `T=20`.

(g) Compare `ρ''` and `ρ'A'`.

2. The following two functions carry out exactly the same operations.

```
      ∇ A COM B                      ∇ A COMP B
[1] →B×⍳A > B                  [1] →4+2××A-B
[2] →S×⍳A < B                  [2] 'LESS'
[3] 'EQUAL'                    [3] →0
[4] →0                         [4] 'EQUAL'
[5] B: 'GREATER'               [5] →0
[6] →0                         [6] 'GREATER'
[7] S: 'SMALLER'               [7] ∇
[8] ∇
```

(a) What must the arguments be—scalars, arrays, or characters?

(b) What is the result of the functions?

3. The following function is meant to calculate the simple correlation coefficient between two variables X and Y.

```
      ∇ R←X SC Y;MX;MY;NUM;DENOM
[1] MX←(+/X)÷ρX
[2] MY←( +/Y )÷ ρ Y
[3] NUM←+/(X-MX)×(Y-MY)
[4] DENOM←((+/(X-MX)*2)×+/(Y-MY)*2
[5] R←NUM÷DENOM
      ∇
```

The mathematical formula for the correlation coefficient is

$$r = \frac{\Sigma_i(X_i - \bar{X})(Y_i - \bar{Y})}{\sqrt{\Sigma_{i,j}(X_i - \bar{X})^2}\sqrt{(Y_j - \bar{Y})^2}}$$

Find the two mistakes in the APL function.

4. The function *BICO* was constructed to calculate the coefficients of the expansion $(a + b)^n$ for given n, i.e., the binomial coefficients. See if you are getting the desired result. If not, can you detect the error?

```
    ∇ R←BICO N
[1] R←1
[2] R←( 0,R)+R,0
[3] →N≥ ρ R/2 ∇
```

5. Write a function that will give you the first N elements of the Fibonacci series 1, 1, 2, 3, 5, 8, 13, . . . (Each number in the series is equal to the sum of the two previous numbers.)
6. My objective was to draw a pyramid. I tried: `' Δ'[1+(ι20)∘.≤ι20]` and I got a tilted pyramid. Can you help me?
7. Write an APL expression that will replace with K all of the elements less than or equal to K in a given array W.
8. The powerful jot dot operator and its uses are summarized in the following exercises.

(a) `X∘.×X←ι20`	(g) `(X+1)∘.⍟X`	(m) `X∘. ⌊⌊/X`
(b) `X∘.=X`	(h) `X∘.*X`	(n) `X∘.≠X`
(c) `X∘. > X`	(i) `X∘.*0`	(o) `X∘.≤10`
(d) `X∘.≥X`	(j) `0∘.*X`	(p) `X∘.≥15`
(e) `X∘. < X`	(k) `1∘.*X`	(q) `X∘.!30`
(f) `X∘.≤X`	(l) `X∘.⌈⌈/X`	(r) `X∘.*2`

9. Show that the expression `+/(ιN)∘.*2` is equivalent to $(1 + N) \times (1 + 2 \times N) \times N \div 6$ for any integer N.
10. The generalized inner product has more uses than we showed in the text. Consider `A←3 4 5    B←ι5`

(a) `A×.-B`	(g) `A!.⌊ /A`
(b) `A×.-2`	(h) `A+.×2`
(c) `2×.-A`	(i) `A×.+2`
(d) `B⌈.-A`	(j) `A+.⍟A`
(e) `B⌈. ⌊B`	(k) `A⍟.*A`
(f) `A!. ⌊ /A`	

Since there are twenty-four primitive (scalar) dyadic functions, there are 406 `(2!  24)` possible combinations. Good luck.

11. Enter the function *TTEST* (pages 135–136) on your workspace and carry out the following tasks:
    (a) Display the function but don't close it.
    (b) Between lines [12] and [13] put the line

```
'THE T-STATISTIC IS'
```

(c) Rename the function as

```
X TTEST Y
```

(d) Put the appropriate local variables in the function header.

Statistical Applications

1. The nicotine content in milligrams of a random sample of two kinds of cigarettes is given below.
Brand A: 16.2 17.7 16.7 15.9 15.1
Brand B: 14.8 17.5 16.1 13.3 15.6
Use your *TTEST* function to find out if the two brands have on the average the same nicotine content. (These are not paired observations.)

2. Ten persons engaged in a prescribed program of physical exercise for weight reduction for a period of one month. The results are given in the following table.

	Weight (lb)									
Before the program:	208	215	196	185	232	156	188	195	232	167
After the program:	205	219	185	175	207	132	195	158	198	167

Was the program effective? The program is effective if the mean weight after the program is statistically smaller than the mean weight before the program. (*Hint:* Let Y = after − before.)

3. In 1960, a sample of 300 ten-year-old boys had mean height 52.8 inches, with standard deviation 2.3 inches. In 1975, a sample of 600 ten-year old boys had mean height of 53.9 inches, with standard deviation 2.5 inches. Since the samples were very large, you may take σ_1^2 and σ_2^2 as being known: $\sigma_1^2 = (2.3)^2 = 5.29$; $\sigma_2^2 = (2.5)^2 = 6.25$ so that $\bar{X}_1 - \bar{X}_2$ has a variance of $(5.29/300) + (6.25/600) = 0.28$. Can we conclude that on the average the ten-year-old boys of the seventies are taller than the ten-year-old boys of the sixties?

4. Here is a group of problems that ask you to check if the Poisson distribution can be used to fit some data. We remind you that the Poisson distribution is given by $P(X) = M^X e^{-M}/X!$, $X = 0, 1, 2, \ldots$ and that
(i) $E(X) = M$
(ii) $VAR\ (X) = M$
(iii) $S_X = \sqrt{M}$
(iv) $a_3 = 1/\sqrt{M}$, coefficient of skewness
(v) $a_4 = 3 + 1/M$, coefficient of kurtosis.
Notice that since $a_3 > 0$, the Poisson distribution is skewed to the right.

Combine the *POISSON* function (page 75), the *MNTS* function (page 72 or 92), the *HIST* function (pages 76–77), the POISSON tables of exercise 8, Chapter 5 (page 93) into a *POISSON FIT* function whose output will consist of

(i) The first four sample moments.

(ii) The sample histogram.

(iii) The theoretical values of X.

(iv) The probabilities of X being greater or less than a specified value given in the function header.

Use this function to solve the following problems:

(a) A study was made to determine whether the deaths of centenarians were distributed randomly over time. In the data below, X represents the number of such deaths that occurred in any one day, and f represents the number of such days in a set of one thousand days.

X	0	1	2	3	4	5	6	7	8	9
f	229	325	257	119	50	17	2	1	0	0

Fit the Poisson distribution to these data. Find the probability that at most one death occurs on a given day.

(b) In the following table, X represents the number of shirts bought by f men that walked into a clothing store in one day.

X	0	1	2	3	4
f	35	40	16	8	1

Use the *POISSON FIT* function to check if the Poisson distribution fits the data satisfactorily. Find the probability that a randomly selected customer will buy at least one shirt.

(c) The following table gives the number of times (X) that your APL terminal malfunctioned, and the respective number of days (f) that a malfunction occurred during a one year period.

X	0	1	2	3	4	5
f	220	80	50	8	2	0

Fit the Poisson distribution to these data and find the probability that during a specific day you will have no problem with your terminal.

5. The owner of a bakery knows that the daily demand for a highly-perishable cheesecake is as shown in the following table.

Daily Demand	Probability
0	0.05
10	0.15
20	0.25
30	0.30
40	0.15
50	0.10
	1.00

Since the most likely daily demand is 30 cakes, he decides to cook 30 cakes per day. Assume that for each cheesecake he cooks and sells he makes a profit of $3, while for each one he cooks but does not sell he loses $2 (assuming that he must throw it away). (*Hint:* For every number he could bake between 1 and 50, use the computer to compute the expected profit.)

(a) Find his expected profit if he decides to cook 30 cakes. (*Hint*: If he cooks 30 and sells 0 he loses $60, and this happens with probability 0.05, and so on.)

(b) Find the optimum number of cakes that he can bake (i.e., the number that will maximize his expected profit). Notice that the profit-maximizing daily demand does not equal the maximum daily demand.

6. A process for making steel pipe is under control if the diameter of the pipe is 5 inches. The known value of the standard deviation is 0.015 inches. In order to test whether the process is under control, a random sample of size 30 is taken, and the mean value of the sample is found to be 3.0078. If a level of significance of $\alpha = 0.01$ is used, should the process be adjusted or not?

9

Elementary Linear Regression, Goodness of Fit, and Analysis of Variance (ANOVA) Problems

9.1 Introduction to Linear Regression

Simple Least Squares Regression

Let us begin with the simplest form of a linear regression model—one found and discussed in every introductory textbook on statistics or econometrics; some useful textbooks are listed in the bibliography at the end of the book. The mathematical model is

$$Y = a + b\,X + U$$

where Y is the regressand, X is the regressor, and U is called a disturbance term. Only Y and X are observed; U is unobserved. All textbooks assume that you have n observations, or, in our new *APL* language, we would say we have two arrays of length n called Y and X.

Estimators of Regression Coefficient

The idea of regression analysis is to find values for a and b, say $\hat{a}$, $\hat{b}$, such that we minimize the sum of squared deviations (differences) between the elements of the array Y and those of the array $\hat{Y}$ defined by $\hat{Y} = \hat{a} + \hat{b}\,X$. Mathematically, this means that $\hat{b}$, $\hat{a}$ are given by

$$\begin{aligned}\hat{b} &= \Sigma_1^n(X_i - \bar{X})\,(Y_i - \bar{Y})/\Sigma(X_i - \bar{X})^2\\ &= (\Sigma\, X_iY_i - n\bar{X}\bar{Y})/(\Sigma\, X_i^2 - n\bar{X}^2)\end{aligned}$$

where

$$\begin{aligned}&\bar{X} = \Sigma\, X_i/n,\ \bar{Y} = \Sigma Y_i/n,\ n\bar{X}\bar{Y} = \Sigma Y_i\bar{X}\\ &\hat{a} = \bar{Y} - \hat{b}\,\bar{X}\end{aligned}$$

Estimators of the Variance of Regression Estimators

Correspondingly, the estimated variance of U and the variances of the coefficient estimators a, b are defined mathematically by

$$\begin{aligned}s^2 = \widehat{\text{VAR}}\,(u) &= \Sigma(Y_i - \hat{Y}_i)^2/(n-2)\\ \widehat{\text{VAR}}\,(\hat{b}) &= s^2(\Sigma(X_i - \bar{X})^2)^{-1}\\ \widehat{\text{VAR}}\,(\hat{a}) &= s^2(n^{-1} + \bar{X}^2/\Sigma(X_i - \bar{X})^2)\\ \widehat{\text{VAR}}\,(\hat{a},\hat{b}) &= -\,\bar{X}s^2/\Sigma(X_i - \bar{X})^2\end{aligned}$$

Breaking up the Sum of Squares

An important identity in regression analysis is that the total sum of squares is equal to the sum of the regression and error sums of squares. Algebraically, this is

$$\underset{\text{SST}}{\Sigma(Y_i - \bar{Y})^2} = \underset{\text{SSR}}{\Sigma(\hat{Y}_i - \bar{Y})^2} + \underset{\text{SSE}}{\Sigma(Y_i - \hat{Y}_i)^2}$$

Here SST is the *total* sum of squares, SSR is the *regression* sum of squares, and SSE is the *error* sum of squares.

The coefficient of determination is given by the formula.

$$R^2 = \text{SSR/SST} = 1.0 - \text{SSE/SST}$$

9.2 An APL Program for Linear Regression Analysis

Once you have a clear and accurate mathematical statement of what you want, you can write down the appropriate APL expression. Let us define a little APL regression function to calculate the above statistics for any pair of arrays Y and X.

We will define a regression routine which does not return an explicit result, which has two arguments, and we will make sure that all other variables used in the routine are local. Now that we know what to do, we can easily write down (on paper) each line in sequence. As we add variables, which we want to be local, in the body of the routine, we can add the variable names to the header. When the routine is written out on a piece of paper, we can check it over for errors and then enter it on the terminal.

The header will start as

```
∇ Y REGRESS X
```

An examination of the series of mathematical formulas listed above shows that the calculation of $\hat{b}$ is central to the whole analysis, so let's begin by defining B, the APL symbol representing the estimator $\hat{b}$. The following series of lines will give you one idea of how to go about writing a "one-liner". Start off with the basic mathematical notion and work outwards from there, keeping in mind that the computer reads from right to left. But do not expect to get everything not only right, but elegant or even computationally efficient, the first time. Let's define B, the APL variable for $\hat{b}$. Each line of the text below represents a *subsequent stage* in fitting together the whole line.

```
[1] B ←   (Y+ .× X) -+/Y× XM
```

XM is going to be the mean of X and is useful to define equations needed later

```
[1] B ← ( Y+ .×X)-( +/Y)×XM÷ + /
( X-XM← ( + /X)÷N)*2
```

N is going to be the length of X and will be used a lot below

```
[1] B←(( Y + .×X)-( +/Y)×XM)÷
(XSQ ← + / ( X-XM← ( +/X)÷N←ρ X)*2)
```

XSQ is the sum of squared deviations of X from XM, which will also be used repeatedly

Reread line [1] from right to left, ensuring that it does what is intended and that the parentheses match.

```
[2] A← + /Y÷N-B ×XM
```

+/Y÷N is inefficient; Y÷N-B× XM does not work as intended

```
[2] A← (( +/Y)÷N)-B×XM
```

This is nearly right, but we will need the mean of Y later

```
[2] A← ( YM← ( +/Y)÷N)-B×XM
```

Check this line, which now seems reasonable, and let's proceed to the next. The remaining lines will give you no problems. In the preface we computed these coefficients using Y⌹X and we could have used that method here too. X would be an N-by-2 array where the first column is composed of 1s. This method is presented in Chapter 13.

The whole program might look like this:

Regression Routine

```
      )CLEAR
CLEAR WS
      ∇ Y REGRESS X
[1]  B←(( Y + .×X)-( +/Y)×XM)÷ (XSQ← +/ (X-XM←( +/X)÷N←ρ X)*2)
[2]  A← ( YM← ( +/Y)÷N)- B×XM
[3]  SST← ( +/ (Y-YM)*2)
[4]  SSE←+ / ( Y- ( A+B×X))*2
[5]  V←SSE÷ ( N-2)
[6]  VB←V÷XSQ
[7]  VA← ((( XM*2)÷XSQ)+÷N)×V
[8]  COV←-XM×V÷XSQ
[9]  RSQ←1-SSE÷SST
[10] 'SAMPLE SIZE IS                  ';N
[11] 'MEAN OF X IS                    ';XM
[12] 'MEAN OF Y IS                    ';YM
[13] 'VARIANCE OF X IS                ';XSQ÷N
[14] 'VARIANCE OF Y IS                ';SST÷N
[15] 'VARIANCE OF ERROR IS            ';V
```

```
[16] 'COEFFICIENT OF A =                    ';A
[17] 'VARIANCE OF A =                       ';VA
[18] 'T-RATIO =                             ';A÷VA*0.5
[19] 'COEFFICIENT OF B =                    ';B
[20] 'VARIANCE OF B =                       ';VB
[21] 'T-RATIO =                             ';B÷VB*0.5
[22] 'COEFFICIENT OF DETERMINATION IS       ';RSQ
[23] 'COVARIANCE  ( A, B ) IS               ';COV
     ∇
     Y←55 70 90 100 90 105 80 110 125 115 130 130
     X←100 90 80 70 70 70 70 65 60 60 55 50
     Y REGRESS X
```

Sample Results of Regression Routine

```
SAMPLE SIZE IS                        12
MEAN OF X IS                          70
MEAN OF Y IS                          100
VARIANCE OF X IS                      187.5
VARIANCE OF Y IS                      525
VARIANCE OF ERROR IS                  69.889
COEFFICIENT OF A =                    210.44
VARIANCE OF A =                       158.03
T-RATIO =                             16.74
COEFFICIENT OF B =                    ¯1.5778
VARIANCE OF B =                       0.031062
T-RATIO =                             ¯8.95
COEFFICIENT OF DETERMINATION IS       0.88907
COVARIANCE  ( A, B ) IS               ¯2.1743
```

Now you know how to calculate various statistics used in simple linear regression.

We can define $\hat{Y} = \hat{a} + \hat{b}X$, the residuals $\hat{U} = Y - \hat{Y}$, plot relationships between them, calculate the variance of the estimator $\hat{Y}$, and so on. Just for practice, try the routine *RESID*, which will calculate the residuals and $\hat{Y}$ ready for plotting or otherwise analyzing the results. Consider

```
∇RESID[⎕]∇
```

Residual Routine

```
     ∇ RESID
[1] YH←A+B×X
[2] RES←Y-YH
[3] 'RESIDUALS'
[4] RES
[5] 'COMPUTED VALUES OF Y→→→→→→Y-HAT'
[6] YH
     ∇
     RESID
RESIDUALS
   2.333333333 1.555555556 5.777777778 0 ¯10 5 ¯20 2.111111111
   9.222222222 ¯0.7777777778 6.333333333 ¯1.555555556
COMPUTED VALUES OF Y→→→→→→Y-HAT
   52.66666667 68.44444444 84.22222222 100 100 100 100 107.88888
   115.7777778 115.7777778 123.6666667 131.5555556
```

9.3 Goodness of Fit, Contingency Tables, and ANOVA Problems

Chi-Square Distribution and Goodness of Fit

Chi-Square and Goodness of Fit

A number of statistical analyses of data are linked by their common reliance on the chi-square distribution. The chi-square distribution is the distribution of the random variable Y, defined by $Y = \Sigma_1^N U_i^2$, where the U_i are independent random variables, each distributed as a standardized normal. That is, it has a mean of zero and a variance of one. Stated symbolically, U_i is distributed as $N(0, 1)$. The chi-square distribution depends only upon its degrees of freedom; in the above example, the distribution of Y is chi-square with N degrees of freedom.

One of the first uses of the chi-square distribution you will encounter in your statistical travels is the ''goodness of fit'' test. In the goodness of fit test you are asked to decide whether an observed set of relative frequencies is consistent with some theoretical predictions. So suppose that you have the traditional k cells; there are a total of N observations, n_1 of which are in the first cell, n_2 in the second, and so on. $n_1 + n_2 + \cdots + n_k = N$. You assume that someone (maybe your favorite uncle) has given you the theoretical probabilities P_i of getting an observation in each of the cells. Thus P_1 is the probability for the first cell, P_2 is the probability for the second, and so on, so that the expected number of entries in each cell is $NP_1, NP_2, \ldots, NP_k$. Your task is to compare the n_i with the NP_i, $i = 1$,

2, . . . , k, and the statistic you calculate is written mathematically as

Chi-Square Distribution

$$W = \sum_1^k \left(\frac{(n_i - NP_i)^2}{NP_i}\right)$$

where W is distributed as a chi-square variable with $(k - 1)$ degrees of freedom.

The APL expression is obtained in a straightforward manner. Our inputs to our chi-square function are two arrays, one for the n_i and one for the P_i, $i = 1, 2, \ldots, k$. So let's begin by entering them:

```
N←15 7 4 11 6 17

PI←6 ρ ÷6
```

We have here an experiment to test whether a die is unbiased. On the assumption that the die is unbiased, $P_i = 1/6, i = 1, 2, \ldots, k$. This is a very simple function, so we can build up our expression from the middle (rather, from the most important part of the expression) as follows. Using paper and pencil, we put down

First Stage: `N-PI×NN←+/N`

Second Stage: `((N-PI×NN←+/N)*2)÷NN×PI`

This will not do, since NN (reading from the right) has not been defined early enough.

Third Stage: `+/((N-PI× NN)*2) ÷ PI×NN← +/N`

A suitable goodness of fit routine would look something like the following:

Goodness of Fit Routine

```
    ∇ N GOODFIT PI
[1] G←+/ ((N-PI×NN)*2)÷PI×NN←+/N
[2] 'CHI-SQ GOODNESS OF FIT STATISTIC IS    ';G
[3] 'WITH '; (ρN)-1; ' DEGREES OF FREEDOM'
[4] ∇
```

Let's try our function. Enter it into the computer, and then call it by

```
    N GOODFIT PI
CHI-SQ GOODNESS OF FIT STATISTIC IS 13.6
WITH 5 DEGREES OF FREEDOM
```

Contingency Tables

Contingency Tables

Now we can move on to something a little more ambitious that introduces some new APL concepts. Contingency tables are generated in the following manner. Suppose that someone takes a sample of people, or machines, or whatever, and then classifies everyone or everything into two categories. For example, people can be classified according to blood type and color of eyes, machines by frequency of breakdown (i.e., once a month, twice, three or more times) and number of defective items produced, and so on. The two-way classification produces a table of cells such as that shown in Table 9.1.

Two-Way Classification

Table 9.1. Two-Way Classification Scheme

		Color of Eyes			
		Bl	Br	Gr	G
Blood Type	A B O				

In this example there are 12 (3 × 4) cells altogether. The entry in each cell is the number of people in the sample who have the designated pair of classification characteristics. For example, the entry in the B row and Br column would be the number of people in the sample with blood type B and eye color brown. Everyone in the sample has to be in one (and only one) of the cells, so if we add up all the cell entries we get the number of people in the sample.* Let n_{ij}, $i = 1, 2, \ldots, r$, $j = 1, 2, \ldots, c$ denote the number of people in the cell (i,j), where r and c are the number of rows and columns, respectively, in the table. The hypothesis to be tested with a contingency table is that the observed entries in any row (or column) differ from the expected number in the row (or column) only by sampling variation. The expected number in the cell (i,j) is given by $(n_{i.}n_{.j})/N$, where $n_{i.}$ is the ith row sum and $n_{.j}$ is the jth column sum, and N is the total number of people in the sample (see Table 9.2).

Table 9.2. Expected Number of Observations in the Two-Way Classification Scheme of Table 9.1.

		Color of Eyes				
		Bl	Br	Gr	G	
Blood Type	A	n_{11}	n_{12}	n_{13}	n_{14}	$n_{1.}$
	B	n_{21}	n_{22}	n_{23}	n_{24}	$n_{2.}$
	O	n_{31}	n_{32}	n_{33}	n_{34}	$n_{3.}$
		$n_{.1}$	$n_{.2}$	$n_{.3}$	$n_{.4}$	N

The chi-square statistic for this problem is

$$W = \sum_{i=1}^{r} \sum_{j=1}^{c} \frac{(n_{ij} - n_{i.}n_{.j}/N)^2}{n_{i.}n_{.j}/N}$$

which, if the assumption is true, is distributed as a chi-square distribution with $(r - 1) \times (c - 1)$ degrees of freedom.

* Our appologies to those with blood group AB or hazel eyes. We simply wanted to keep the size of the table down.

Dyadic ρ: Reshape

Reshape ρ

Our first problem in handling this situation is how to enter a table like the one above into the computer. This reintroduces the dyadic use of ρ, called the "reshape" function. Consider the following examples:

```
      A←2 3 ρ 1 2
      A
1 2 1
2 1 2
      B←2 2 ρ1 2 3 4
      B
1 2
3 4
      C←2 2 ρ 1 2 3 4 5 6
      C
1 2
3 4
```

The use of reshape rearranges the array given on the right-hand side into the shape denoted by the left-hand side. If D is an array, then E←M N ρ D will take the first n elements of D and put them down as the *first row* of E, then the second n elements of D become the *second row* of E, and so on, m times. If there are too few elements in D, as shown in the example with A, then the elements of D are reused from the beginning until a table of appropriate size is created. In APL such tables are known as two-dimensional arrays or tables, as opposed to the one-dimensional arrays or lists we have been having fun with so far.

Our first task is to obtain the row and column sums. With a one-dimensional array we know how to do this: +/(plus reduction). But with rows and columns, what happens? Try

```
      +/A
4 5
```

Plus Reduction of Rows and Columns

So +/ on a two-dimensional array gives us plus reduction of the *rows* of A. We get plus reduction of columns by a simple device. Try overstriking the reduction sign with the subtract sign; thus:

```
      +⌿A
3 3 3
```

gives us plus reduction down columns.

Outer Product

Outer Product ∘.×

Now that we see how to get row and column sums, the next step is to create the table of entries: $(n_{i.}n_{.j})/N$. This is most easily accomplished by means of what in APL is called the *outer product*. Suppose that x and y are any two single-dimensional arrays, say m and n in dimension, respectively, and we want the table created by multiplying each element of x by each element of y. If x has elements $x_1, x_2, \ldots, x_m$ and y has elements $y_1, y_2, \ldots, y_n$, then we want the $(m \times n)$ table of entries given by $x_1y_1, x_1y_2, x_1y_3, \ldots, x_my_n$ arranged into an $(m \times n)$ table. This sounds like a very complicated arrangement, but once again what looks difficult is easy in APL. The $(m \times n)$ table we want is obtained by using the "outer product," or jot dot product, which is ∘.× and is keyed by upper case J (gives the jot), period (gives the dot), and the multiply key. Thus we have

```
T←X∘.×Y
```

and T is the desired $(m \times n)$ table of entries. Let's try it. Recall the X and Y that we have used before. In case you logged-off since last using them, we have

```
      X←1 2 ¯3 ¯4 5 ¯6 ¯7
      Y←2 4 6 8 4 2 6
      T←X∘.×Y
      T
  2   4   6   8   4   2   6
  4   8  12  16   8   4  12
 ¯6 ¯12 ¯18 ¯24 ¯12  ¯6 ¯18
 ¯8 ¯16 ¯24 ¯32 ¯16  ¯8 ¯24
 10  20  30  40  20  10  30
¯12 ¯24 ¯36 ¯48 ¯24 ¯12 ¯36
¯14 ¯28 ¯42 ¯56 ¯28 ¯14 ¯42
```

If you examine the entries of T you will see that the entry in the (i, j) position is the product of the ith element of X and the jth element of Y.

Now we can create a table of entries where the (i, j) entry in the table is $(n_{i.}n_{.j})/N$. Let TB be the two-dimensional array of numbers n_{ij}. Write down our first attempt at creating the table:

```
R←+/TB
C←+⌿TB
TB←TB-R∘.×C÷N←+/R
```

Line [1] gives the row sums and line [2] the column sums. The third line obtains N by adding up row sums, and divides that into the array of column totals which, in turn, is multiplied element by element by the elements of the array *R*. The new array is then subtracted from the original array *TB*, element by element, and the result is stored in *TB* again to save space (the storing of a number of big tables soon uses up all of your available workspace). The new contents of *TB* contain elements such as $(n_{ij}-n_{i.}n_{.j}/N)$.

The elements of the new array *TB* have to be squared and divided by the elements of `(R∘.×(÷N←+/R)`. Let's try again.

```
      NT←R∘.×C÷N←+/R
      +/+/((TB←TB-NT)*2)÷NT
```

To be sure that you are following all this step by step, let's actually perform these operations with *TB*.

```
      TB←2 3 ρ2 3 1 1 4 2
      R←+/TB
      R
6 7
      C←+⌿TB
      C
3 7 3
      NT←R∘.×C÷N← +/R
      NT
1.384615358      3.230769231      1.384615385
1.615384615      3.769230769      1.615384615
```

A routine for calculating the statistic of a contingency table might look like the following:

Chi-Square Contingency Table Routine

```
      ∇ CONTAB TB;R;C;NT;W
[1]  R←+/TB
[2]  C←+⌿TB
[3]  DF←(ρTB)-1
[4]  NT←R∘.×C÷N←+/R
[5]  W←+/+/(( TB-NT)*2)÷NT
[6]  'THE CHI-SQ STATISTIC FOR A'
[7]  'CONTINGENCY TABLE IS  ';W
[8]  'WITH ';DF[1]×DF[2];' DEGREES OF FREEDOM'
[9]  ∇
```

This routine does not produce a specific result, and variables used as intermediate output are made local to the function. Note that this saves room in the limited workspace available to you. Otherwise, the intermediate products would sit around until you reused them or erased them. Let's try it. Type in

```
    TB←3 4 ρ18 29 70 115 17 28 30 41 11 10 11 20
    CONTAB TB
THE CHI-SQ STATISTIC FOR A
CONTINGENCY TABLE IS 19.94264769
WITH 6 DEGREES OF FREEDOM
```

One-way Analysis of Variance

So much for testing hypotheses about contingency tables. The last type of statistical analysis we will handle in this chapter is analysis of variance (abbreviated ANOVA). Only the simplest types of problems will be dealt with here. The analysis that follows is called *one-way* analysis of variance.

One-way ANOVA

Once again imagine that we have a table of entries. Each column of entries might represent, for example, crop yields with a given type of fertilizer on different types of land, or each column might represent typing speeds of various typists on a variety of machines.

If x_{ij} represents the entry in the (i, j)th position, for example, the ith typist using the jth machine, then $\bar{x}_{.1}, \bar{x}_{.2}, \ldots, \bar{x}_{.c}$ represent the mean typing speeds of all typists using the first machine, the second machine, and so on up to the cth machine—there being c machines and c columns. Let $\bar{x}$ represent the overall mean, i.e., the mean of the column means.

The main concept in analysis of variance is to examine the breakup of the total sum of squares just as we did in the regression section. Thus we consider the identity

$$\underset{\text{TSS}}{\Sigma_{i=1}^{r}\Sigma_{j=1}^{c}(x_{ij} - \bar{x})^2} = \underset{\text{ESS}}{\Sigma_{i=1}^{r}\Sigma_{j=1}^{c}(x_{ij} - \bar{x}_{.j})^2} + \underset{\text{CMSS}}{\Sigma_{i=1}^{r}\Sigma_{j=1}^{c}(\bar{x}_{.j} - \bar{x})^2}$$

where TSS = total sum of squares, ESS = error sum of squares, and CMSS = column mean sum of squares. CMSS can be rewritten as $\Sigma_{i=1}^{r}\Sigma_{j=1}^{c}(\bar{x}_{.j} - \bar{x})^2 = r\Sigma_{j=1}^{c}(\bar{x}_{.j} - \bar{x})^2$.

F - Statistic

The test of the hypothesis that there is no difference in the column means is obtained by looking at the ratio of CMSS$/(c - 1)$ to ESS$/(rc - c)$ which, under the null hypothesis of no difference, is distributed as F with $(c - 1)$ and $c(r - 1)$ degrees of freedom. This is fully explained in statistics books; see, for example, Mendenhall and Reimuth, which is listed in the bibliography. Consequently, we want to calculate the statistic F defined by

$$F = \frac{\text{CMSS}[c(r-1)]}{\text{ESS}(c-1)} = \frac{rc(r-1)}{(c-1)} \frac{\Sigma_{j=1}^{c}(\bar{x}_{.j} - \bar{x})^2}{\Sigma_i\Sigma_j(x_{ij} - x_{.j})^2}$$

Let us suppose we have a table X of c columns of r entries each, and we want to do a one-way analysis of variance on X by columns. From the mathematical expression above we need column sums and the overall sum; but this is now very easy! Write down the following first attempt:

```
MCOL←( +⌿X)÷R←( ρ X)[1]

M←( +/MCOL)÷C←( ρ X)[2]

NUM←R×C×( R-1)×+/( MCOL-M)*2

DEN←( C-1)×+/+/ ( X- (R,C) ρMCOL)*2

F←NUM÷DEN
```

Index of an Array of Dimensions

The only new element here is (ρX)[1] and (ρX)[2]. The expansion (ρX) is an array of the dimensions of X, so that (ρX)[2] gives the second. If the parentheses had been left off, any attempt to execute ρX[1] would have given an error; thus

```
      ρ X [1]
RANK ERROR
      ρ X [1]
         ∧
```

The reason for this is that once X has shape $r \times c$, the expression X[1] is invalid. What is needed to index a *two*-dimensional array is a *pair* of indexing numbers, e.g., X[1;2]. More of this later. What we wanted, of course, was the first element of the *one*-dimensional array (ρX).

You should also recall that to obtain an $(R \times C)$ table from *MCOL*, an array of C elements, we must write (R,C) ρMCOL and not just R C ρMCOL, which produces a SYNTAX ERROR. This is because the pair of variables R C is not an array even though 8 3 would be. To produce an array with a set of variables it is necessary to catenate them; thus, (R,C) is an array of two elements, R and C.

An example of an analysis of variance routine for this simple problem is

One-way ANOVA ANOVA1

```
      ∇ ANOVA1 X
[1] MCOL←(+⌿X)÷R←(ρ X)[1]
[2] M←(+/MCOL)÷C←(ρ X)[2]
[3] NUM←R×C× (R-1)×+/ (MCOL-M)*2
[4] DEN← (C-1)×+/+/(X-(R,C) ρ MCOL)*2
[5] F←NUM÷DEN
[6]  'THE F STATISTIC FOR ANALYSIS OF'
[7] 'VARIANCE ACROSS COLUMNS IS ';F
[8] 'WITH '; (C-1), 'AND ';C×(R-1);' DEG FREEDOM '
[9] ∇
```

Now that we have it, let's try it. Type

```
      TAB←8 3⍴ 44 40 54 39 37 50 33 28 40,⎕
⎕:
      56 53 55 43 38 45 56 51 66 47 45 49,⎕
⎕:
      58 60 65
      ANOVA1 TAB
THE F STATISTIC FOR ANALYSIS OF
VARIANCE ACROSS COLUMNS IS 1.868644068
WITH 2 AND 21 DEG FREEDOM
```

Two-way Analysis of Variance

Two-way ANOVA

The last problem that we will consider in this section is two-way analysis of variance. As before, we have a table of entries with r rows and c columns. With *one-way* analysis of variance we calculated only column means; in *two-way* analysis we calculate both *column and row* means. As before, the analysis of the test of the null hypothesis of "no effects" is based on the breakup of the total sum of squares. Consider the identity:

$$\underset{\text{TSS}}{\Sigma_{i=1}^{r}\Sigma_{j=1}^{c}(x_{ij} - \bar{x})^2} = \underset{\text{ESS}}{\Sigma_{i=1}^{r}\Sigma_{j=1}^{c}(x_{ij} - x_{.j} - \bar{x}_{.i} + \bar{x})^2} + \underset{\text{RMSS}}{\Sigma_{i=1}^{r}\Sigma_{j=1}^{c}(\bar{x}_{i.} - \bar{x})^2}$$

$$+ \underset{\text{CMSS}}{\Sigma_{i=1}^{r}\Sigma_{j=1}^{c}(x_{.j} - \bar{x})^2}$$

$\bar{x}$ is the overall mean, $\bar{x}_{i.}$ is the mean of the ith row, and $\bar{x}_{.j}$ is the mean of the jth column. RMSS is the row mean sum of squares and CMSS is the column mean sum of squares. CMSS can be rewritten (as before) as $r\Sigma_{j=1}^{c}(\bar{x}_{.j} - \bar{x})^2$ and RMSS as $c\Sigma_{i=1}^{r}(\bar{x}_{i.} - \bar{x})^2$.

With these statistics we can test three different hypotheses:

H_1: Row means are equal, column means are unspecified.

H_2: Column means are equal, row means are unspecified.

H_3: Row and column means are both equal (not necessarily to each other).

Hypothesis H_i is tested by the statistic F_i defined mathematically by:

$$F_1 = \frac{\text{RMSS}/(r-1)}{\text{ESS}/[(r-1)(c-1)]}$$

$$F_2 = \frac{\text{CMSS}/(c-1)}{\text{ESS}/[(r-1)(c-1)]}$$

$$F_3 = \frac{(\text{RMSS} + \text{CMSS})/[(r - 1) + (c - 1)]}{\text{ESS}/[(r - 1)(c - 1)]}$$

Under the null hypothesis of no row or column effects, F_1 is distributed as F with $(r - 1)$ and $(r - 1)(c - 1)$ degrees of freedom, F_2 as F with $(c - 1)$ and $(r - 1)(c - 1)$ degrees of freedom, and F_3 as F with $(r - 1) + (c - 1)$ and $(r - 1)(c - 1)$ degrees of freedom.

All we have to do now is to program a routine for calculating the statistics required for a two-way analysis of variance. Let's modify ANOVA1 and call it ANOVA2.

So far we have not paid much attention to trying to make our routines computationally efficient or compact in terms of size of workspace used. This is because we felt you had enough to do in learning statistics and the basics of APL all at once. But, since ANOVA2 will be similar to ANOVA1, let's take the opportunity to do things a little more efficiently than before. To this end, let's expand and simplify algebraically each of the sums of squares listed above.

$$\text{ESS: } \Sigma_{i=1}^{r}\Sigma_{j=1}^{c}[x_{ij}^2 + \bar{x}_{i.}^2 + \bar{x}_{.j}^2 + \bar{x}^2 + 2\bar{x}_{ij}(-\bar{x}_{i.} - \bar{x}_{.j} + \bar{x}) - 2\bar{x}_{i.}(-\bar{x}_{.j} + \bar{x}) - 2\bar{x}_{.j}\bar{x}]$$

$$= \Sigma_{i=1}^{r}\Sigma_{j=1}^{c}x_{ij}^2 - c\Sigma_i\bar{x}_{i.}^2 - r\Sigma_j\bar{x}_{.j}^2 + rc\bar{x}^2$$

$$\text{RMSS: } \Sigma_{i=1}^{r}\Sigma_{j=1}^{c}[\bar{x}_{i.}^2 + \bar{x}^2 - 2\bar{x}_{i.}\bar{x}]$$

$$= c\Sigma_i\bar{x}_{i.}^2 - rc\bar{x}^2 = c[\Sigma_i\bar{x}_{i.}^2 - r\bar{x}^2]$$

$$\text{CMSS: } \Sigma_{i=1}^{r}\Sigma_{j=1}^{c}[\bar{x}_{.j}^2 + \bar{x}^2 - 2\bar{x}_{.j}\bar{x}]$$

$$= r\Sigma_j\bar{x}_{.j}^2 - rc\bar{x}^2 = r[\Sigma_j\bar{x}_{.j}^2 - c\bar{x}^2]$$

From the above expressions we see that the major sums needed are $\Sigma\Sigma x_{ij}^2$, $\Sigma_i\bar{x}_{i.}^2$, $\Sigma_j\bar{x}_{.j}^2$, and $\bar{x}^2$. We might begin by putting together the various sums and parameters required to calculate the main results. Consider, for a table X of entries

```
R←(ρ X)[1]        } Gives us the R, C values
C←(ρ X)[2]        } needed in the calculations

XR←+/X            } Gives the row and column
XC←+⌿X            } sums
```

As noted above, there are three possible F values we can calculate, all of which require ESS/$[(r - 1)(c - 1)]$ in the denominator. Consider the following possible form for ANOVA2:

ANOVA2

```
    ∇ ANOVA2 X;RM;CM
[1]   R←(ρ X)[1]
[2]   C←(ρ X)[2]
```

```
[3]  RM←( +/X)÷C
[4]  GM←( +/ ( CM← ( +⌿X)÷R))÷C
[5]  RMSS←C×((RRM←+/( RM⋆2))-R×GM⋆2)
[6]  CMSS←R×( CCM←+/CM⋆2)-C×GM⋆2
[7]  ESS←(+/+/X⋆2)+(( -C×RRM)+( -R×CCM))+C×R×GM⋆2
[8]  DEN←ESS÷DDF←(( R-1)×( C-1))
[9]  F1←RMSS÷DEN×( R-1)
[10] F2←CMSS÷DEN×( C-1)
[11] F3←( RMSS+CMSS)÷DEN×( R+C-2)
[12] 'THE F STATISTICS FOR TWO-WAY ANOVA ARE: '
[13] 'TESTING ROW MEANS, F IS  ';F1
[14] 'WITH '; ( R-1),DDF;' DEG. FREEDOM'
[15] 'TESTING COL. MEANS, F IS  ';F2
[16] 'WITH '; (C-1),DDF;' DEG. FREEDOM'
[17] 'TESTING ROW AND COL. MEANS, F IS  ';F3
[18] 'WITH '; ( R+C-2),DDF;' DEG. FREEDOM'
[19] ∇
```

There are a few points in this routine worth noting. First, the number of operations performed has been reduced. This was made possible by showing that the required algebraic expressions depended upon a small number of partial sums. Second, intermediate results are stored in a variable for use later in the routine (see, for example, lines [4], [5], and [6]). Third, an attempt was made to ensure that where a sum of variables is divided (or multiplied) by the same constant, the sum is taken first, then the result is divided (or multiplied). In general, it is better to multiply before dividing in a complicated expression since this reduces the propagation of errors introduced by the process of division. This rule was not followed entirely in this routine since it was, in fact, more efficient to form the denominator first, because it was used repeatedly.

Line [7] is interesting in that it poses a trap for those who forget that the computer operates from right to left. If we had written `ESS←+/+/X⋆2-C×RRM-R×CCM+C×R×GM`, we would have obtained an erroneous result. `C×R×GM` is alright, but then this is added to *CCM, which sum* is then multiplied by *R*, leading to the first error. The result (of `R×CCM×C×R×GM`) is subtracted from *RRM*, leading to the second error, and so on. Note that the minus operator signs used with *R* and *C* are being used in their *monadic* sense; $-C$ and $-R$ change the signs of *C* and *R*.

```
      TAB
44 40 54
39 37 50
33 28 40
56 53 55
43 38 45
56 51 66
47 45 49
58 60 65
      ANOVA2 TAB
THE F STATISTICS FOR TWO-WAY ANOVA ARE:
TESTING ROW MEANS, F IS 29.46666667
WITH 7 14 DEG. FREEDOM
TESTING COL. MEANS, F IS 19.6
WITH 2 14 DEG. FREEDOM
TESTING ROW AND COL. MEANS, F IS 27.27407407
WITH 9 14 DEG. FREEDOM
```

Results from ANOVA2

While this routine is neither as compact nor as elegant as it might be, it has served the purpose of demonstrating some of the factors you should be beginning to take into consideration. The major ones are:

1. Keep the use of big arrays, especially tables, to a minimum—they use up an incredible amount of workspace.
2. Simplify the algebra as much as possible *before* beginning to write the routine.
3. Avoid repeating calculations.

9.4 * Calculating the Chi-Square and *F* Distributions

In this short section we will discuss how to calculate the chi-square and F probability distributions so that you need not use the tables and can pick confidence levels and sizes of tests not available there. More importantly, by plotting the distributions you can acquire a better understanding of them than would otherwise be the case.

Let's begin with the chi-square distribution. If Y is distributed as chi-square with N degrees of freedom, its density can be written as:

$$F(Y|N) = \frac{Y^{(N/2-1)}\,\text{Exp}\,(-Y/2)}{\Gamma(N/2)\;2^{(N/2)}}$$

(See, for example, Press in the bibliography.)

Γ Gamma Function
`! X`

The non-APL symbol $\Gamma(N/2)$ represents the gamma function, which generalizes the factorial. If $N/2$ is an integer, then $\Gamma(N/2) = (N/2 - 1)!$. $\Gamma(p)$ is defined by the integral

$$\Gamma(p) = \int_0^\infty x^{p-1}e^{-x}dx$$

If p is an integer, $\Gamma(p) = (p - 1)!$. The chi-square distribution of Y with $N/2$ degrees of freedom is nothing more than the gamma distribution of the variable $Y/2$ with parameter $p = N/2$. In APL, $\Gamma(N/2)$ is easily calculated; in fact, $\Gamma(N/2)$ is given by $!(N \div 2) - 1$, even where $N \div 2$ is a noninteger.

In order to calculate the integral of $F(Y|N)$ from zero to some bound B, we will have to use a numerical approximation. The simplest procedure to follow is that used in obtaining the Normal integral whereby we broke the interval over which the function is to be integrated into a series of small intervals and added up the approximate areas to get the integral. Before continuing you might wish to review briefly section 6.3 in Chapter 6.

How many intervals are needed? Or rather, how small should each interval be? If we regard the interval from 0 to $10\sqrt{2N}$, where $\sqrt{2N}$ is the standard deviation, as the effective range of the variable Y, i.e., integration of the density function over this range gives a value close to one, and if we regard a suitable interval length as a 32nd of a standard deviation, then integration over the effective range would involve 320 intervals. For an integration from 0 to B, we need to determine how many intervals are needed. Note that we do not want to divide the interval 0 to B by some fixed number since the accuracy of the integral would vary tremendously with variations in the value of B. Figure 9.1 will help clarify our stratagem. In the sample shown, the interval zero to B should be broken up into 9 sub-intervals.

Figure 9.1 Illustration of a Method for Determining Integration Intervals

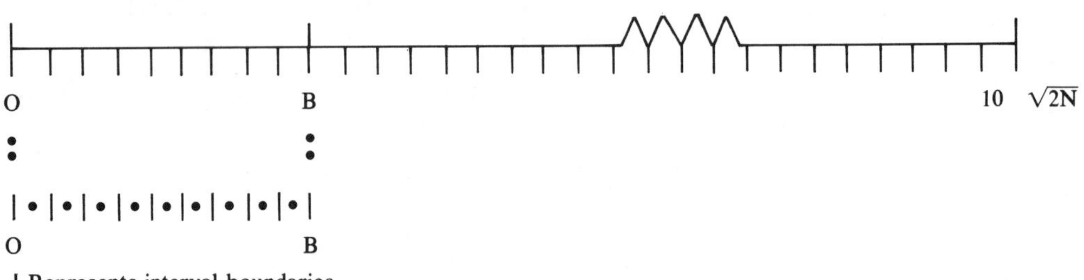

Thus, the required number of sub-intervals can be obtained in APL as follows:

```
L← (0,⍳320)×((2×N)*0.5)÷32

NINT←+ /B>L
```

The first line produces 321 points on the interval from 0 to $10\sqrt{2N}$ and the second line gives, by compression, the numbers of such points less than B and hence the number of intervals into which 0 to B is to be divided.

The idea underlying the approximation of the integral is uncomplicated.* On each sub-interval, we will approximate the actual area under $F(Y|N)$ by a rectangle whose base is given by BW $(B - 0) \div$ NINT and whose height is $F(Y^*|N)$, where Y^* is the mid-point of the interval. This approximating procedure is known as the rectangular method. It is the same as that used in Chapter 6 to calculate the cumulative normal distribution. A brief glance at figure 6.2b will remind you of the main idea.

The sequence of mid-points at which $F(Y|N)$ is to be calculated is given by

```
YS←((⍳NINT)×BW)-BW÷2
```

In Figure 9.1 above, these points are shown as the dots on the interval $[0, B]$.

We now have most of the pieces, so that all we have to do is fit them together. Consider

Cumulative Chi-Square

```
      ∇ W←N CUMCHISQ B;YS                      ←N = degrees of freedom
                                                 B = boundary of integration
[1]   DEN←(2*NN)×!(NN←N÷2)-1                   ←Calculates denominator
                                                 of F(Y|N)
[2]   NINT←+/B > L←(0,⍳320)                    ←{Determines number of
      ×((2×N)*0.5)÷32                            {intervals needed
[3]   YS←(( ⍳NINT)×BW)-(BW←B÷NINT)÷2 ←Midpoints of the intervals
[4]   FYS←( YS*(NN-1))×*(-YS÷2)                ←Evaluates the density at YS
[5]   W←(W←+/FYS×BW)÷DEN                       ←{W gives probability of
[6]   'THE CUM CHI SQ DISTN'                     {random variable distributed
[7]   'WITH ';N;' DEG FREEDOM'                   {as F(Y|N) of being less
                                                 {than B.
[8]   'F(B) IS ';W
[9]   '1-F(B) IS ';(1-W)
[10]  ∇
```

Let's try our function and compare its results with those obtained from the Biometrika Tables prepared by Pearson and Hartley. The official tables give the following numbers.

Comparison of APL-STAT Results and Tabulated Values

Biometrika Table Chi-Sq Values

D.F. \ B	0.1	1.0	5.0	10.0
1	0.24817	0.68269	0.97465	0.99943
2	0.04877	0.39347	0.91791	0.99326
5	0.00016	0.03743	0.58412	0.92476
10	—	0.00017	0.10882	0.55951
20	—	—	0.00028	0.03183

* Any advanced calculus text or book on numerical approximation will discuss various methods. One useful reference is W. Kaplan, *Advanced Calculus*, Addison-Wesley, 1952, Reading, Mass.

Approximate APL Chi-Sq Values

D.F. \ B	0.1	1.0	5.0	10.0
1	.20434	.63228	.92400	.94767
2	.04877	.39345	.91788	.99322
5	.00016	.03741	.58414	.92478
10	—	.00017	.10881	.55951
20	—	—	.00028	.03182

There are a few items worth mentioning with this routine. First of all, in the header to the function, the name was given as *CUMCHISQ*, instead of something like *CUMCHI-SQ*. The reason for this is that a symbol such as "−" cannot be used in the definition of a function name. Further, and very importantly, the way in which the function is written, the arguments N and B must be *scalars, not arrays*. If you define either N or B as an array and attempt to use *CUMCHISQ*, you will get the response

```
LENGTH ERROR
CUMCHISQ[1] NINT←+/B > L←(0,320)×(2×N)*0.5)÷32
                                  ∧
```

if N is an array, or the caret will be under B, if only B is an array.

However, the most important lesson from the above tables is that while the routine provides a useful approximation to the correct values for degrees of freedom greater than 1, the approximation is very bad for one degree of freedom, and we shall not bother with this now. However, you should note that the difficulty is not simply one in which the interval width is too broad. If you were to examine a plot of the densities for 1, 2, 3, and 4 degrees of freedom, you would see considerable differences in the shape of the chi-square distribution as the degrees of freedom increase in this range as shown in the graph in problem 10. What is required is a much more accurate approximating procedure—our simple rectangular procedure is inadequate for the task; some suggestions are contained in the exercises.

Let us consider the F distribution with k_1 and k_2 degrees of freedom. The density function can be written as:

$$f(w) = \frac{(k_1/k_2)^{k_1/2}\Gamma((k_1 + k_2)/2)}{\Gamma k_1/2)\Gamma(k_2/2)} w^{(k_1/2)-1}(1 + (k_1/k_2)w)^{-(k_1+k_2)/2}$$

The terms involved in the constant of integration present no difficulty after the chi-square distribution. Since our strategy for obtaining the cumulative distribution worked reasonably well with the chi-square distribution, let's try it with the F distribution, whose standard deviation in algebraic terms is:

$$\frac{2k_2^2(k_1 + k_2 - 2)}{k_1(k_2 - 2)^2(k_2 - 4)}$$

where k must be greater than 4 for the variance to be defined.

Cumulative F Distribution Routine

The F cumulative distribution function routine can be set up in a manner very similar to that of the chi-square distribution. The main differences occur in the fact that the F distribution has two degrees of freedom instead of one and "FYS" has a different definition in the two routines. Try the following:

```
      ∇ W←P CUMF B
[1]   ⍝THIS ROUTINE CANNOT BE USED WITH
[2]   ⍝LESS THAN FIVE DENOM, DEG. OF FREEDOM.
[3]   CONST←(!((PS←+/P)÷2)-1)×P[1]*P1←P[1]÷2
[4]   CONST←CONST÷(!(P[2]÷2)-1)×(!P1-1)×P[2]*P1
[5]   STD←2×(PS-2)×P[2]*2
[6]   STD←(STD÷(P[1]×(P[2]-4)×(P[2]-2)*2))*0.5
[7]   NINT←+/B > L←(0,⍳320)×STD÷32
[8]   YS←(( ⍳NINT)×BW)-(BW←B÷NINT)÷2
[9]   FYS←(YS*P1-1)×(1+(P[1]÷P[2])×YS)*( -PS÷2)
[10]   W←(W←+/FYS×BW)×CONST
[11]   'THE CUMULATIVE F DISTN WITH'
[12]   P[1];' AND ';P[2];' DEG. FREEDOM'
[13]   'IS ';W
[14]   ∇
```

Comment ⍝

The first two lines of this routine form a comment which will warn the user when he displays the function that it cannot be used with less than five denominator degrees of freedom. The comment symbol is made by keying uppershift C (called cap), backspace, and uppershift J (or jot). Everything to the right of this symbol is regarded as a comment. When the function is executed, comments are ignored; *they are only printed when the function is displayed.*

B is one limit of integration; 0 is assumed to be the other, in that we are calculating the probability of the random variable being contained in the interval $[0, B]$. The array P contains the degrees of freedom, k_1, k_2.

In the calculation of the integral of $f(w)$ there are two component parts multiplied together, a constant:

$$\frac{(k_1/k_2)^{k_1/2}\Gamma((k_1 + k_2)/2)}{\Gamma(k_1/k_2)\Gamma(k_2/2)}$$

and that portion of the function to be integrated is

$$w^{(k_1/2)-1}(1 + (k_1/k_2)w)^{-(k_1+k_2)/2}$$

In calculating the integral the multiplicative constant can be calculated separately from the second part and the two multiplied together to get the answer.

Lines [3] and [4] determine the value of the constant term; line [3] produces the numerator and [4] the denominator. Remember that $\Gamma(k_1/2)$ is obtained in APL by `!(P[1]÷2)-1` since algebraically, $\Gamma(k_1/2) = (k_1/2 - 1)!$, when $k_1/2$ is an integer. The APL function "factorial", !, is really the gamma function.

Lines [5] and [6] calculate the standard deviation of the distribution which is needed for determining the interval widths.

Lines [7] to [10] determine the probability of the random variable lying in the interval [0, B] in a manner very similar to the previous effort.

As a check on the accuracy of the routine, examine the following table. The table shows the probabilities obtained from the routine using various combinations of degrees of freedom and bounds. At the head of each column the theoretically correct probability is listed.

Table of Probabilities from *F* Distributions*

D.F. \ Prob.	.75	.95	.99	.999
(1,5)	0.66 (1.69)	0.86 (6.61)	.90 (16.26)	0.91 (47.18)
(1,10)	0.69 (1.49)	0.89 (4.96)	0.93 (10.04)	.94 (21.04)
(2,5)	0.75 (1.85)	0.95 (5.79)	0.99 (13.27)	0.998 (37.12)
(4,5)	0.75 (1.89)	0.95 (5.09)	0.99 (11.39)	1.000 (31.09)
(4,10)	0.75 (1.59)	0.95 (3.48)	0.99 (5.99)	0.999 (11.28)
(10,20)	0.75 (1.40)	0.95 (2.35)	0.99 (3.37)	0.999 (5.08)

* The bound B is given in parentheses below the probability.

Once again, as with the chi-square distribution, we see that one degree of freedom in the numerator causes difficulties with the approximation. As both degrees of freedom increase, the relative accuracy increases. This is because the functions being approximated are easier to handle with our simple rectangular procedure. Greater accuracy and less computation can be obtained with a more sophisticated procedure, such as the use of Simpson's rule.

Example of Use of Cumulative F Distribution Routine

An example of use of the routine is:

```
      P
1 5
      B
1.69
      P CUMF B
```

```
THE CUMULATIVE F DISTN WITH
1 AND 5 DEG. FREEDOM
IS 0.662880113
0.662880113
```

Summary

REGRESS: Simple least squares regression—an APL routine was provided to calculate most of the relevant statistics in a simple linear regression model of the type $Y = a + bX + U$.

GOODFIT: a routine to perform goodness of fit tests.

CONTAB: a routine for calculating the statistics required in tests of hypotheses within a contingency table.

ANOVA1: a routine for carrying out simple one-way analysis of variance.

ANOVA2: a routine for performing two-way analysis of variance.

Reshape, ⍴ (uppershift *R*): a dyadic function which rearranges the "array shape" of the right-hand argument as specified by the left-hand array. Example:

```
C←2 3 ⍴A
```

rearranges the elements of the array A into a table (matrix) composed of two rows of three columns each; elements from A are stored in the variable C *row* by *row* in sequence from the elements of A.

Outer Product, ∘.× (jot dot) (uppershift *J*), period, multiplication): a dyadic function which multiplies each of the m elements in the left-hand argument array with each of the n elements in the right-hand argument array to form a table of dimension $(m \times n)$ containing all $m \times n$ multiplications.

Reduction, /: a monadic function. When used over two-dimensional arrays (tables or matrices) *f*/*TABLE* produces an array formed by the *f* reduction of each *row* of *TABLE*, where *f* is one of the arithmetic functions.

Reduction, / by *columns* for a two-dimensional array is obtained by using ⌿ (reduction, backspace, minus sign).

Gamma function, ! (uppershift *K*, backspace, period): a monadic function to evaluate the gamma function, which generalizes the factorial function. !*P* in APL produces the mathematical result of $\Gamma(P + 1)$.

CUMCHISQ: a routine to calculate the integral of the chi-squared density function.

CUMF: a routine to calculate the integral of the F density function.

Comment, ⍝ (uppershift C, backspace, uppershift J) used as the first character of a line inside a function to provide explanatory comments when function is displayed. Comments are not printed on execution of function and are otherwise ignored.

Exercises

APL Practice

1. For the two arrays `X←⍳20` and `Y←¯5+⍳20`, use your knowledge of APL to perform the following calculations based on the mathematical formulas listed below. These formulas are useful in regression analysis.

(a) $B = \dfrac{\Sigma_1^n(x_i - \bar{x})(y_i - \bar{y})}{(x_i - \bar{x})^2}$ — estimator of the regression slope coefficient. $\bar{x} = \Sigma x_i/n$ and $\bar{y} = \Sigma y_i/n$.

(b) $A = \bar{y} - B\bar{x}$ — estimator of the constant term

(c) $u_i = y_i - A - Bx_i, \quad i = 1, 2, \ldots, n$ — vector of estimated errors

(d) $\Sigma_1^n u^2 = \Sigma_1^n(y_i - A - By_i)^2$ — sum of the squares of the error.

(e) $S_u^2 = \dfrac{\Sigma_1^n u^2}{n - 2}$, — estimator of variance of disturbance terms

(f) $S_b^2 = \dfrac{S_u^2}{\Sigma(x_i - \bar{x})^2}$ — estimator of the variance of the slope

(g) $S_a^2 = S_u^2 \left(\dfrac{1}{n} + \dfrac{\bar{x}^2}{\Sigma(x_i - \bar{x})^2}\right)$ — estimator of the variance of the constant term

(h) $\mathrm{COV}\,(A, B) = \left[\dfrac{-\bar{x}^2}{\Sigma(x_i - \bar{x})^2}\right] S_u^2$ — estimator of the variance of the covariance of the estimators A and B

(i) $R = \dfrac{\Sigma(x_i - \bar{x})(y_i - \bar{y})}{nS_xS_y}$

where

$S_x = (\Sigma(x_i - \bar{x})^2/(n - 1))^{1/2}$, and
$S_y = (\Sigma(y_i - \bar{y})^2/(n - 1))^{1/2}$ — simple correlation coefficient of x and y

(j) $\mathrm{SST} = \Sigma(y_i - \bar{y})^2$ — total sum of squares
$\mathrm{SSR} = ((A + Bx_i) - \bar{y})^2$ — regression sum of squares

(k) $\mathrm{RSQ} = \dfrac{\mathrm{SSR}}{\mathrm{SST}}$ and compare RSQ to R^2 which is calculated using R from (i).

(l) $V(y|x_0) = S_u^2 \left[\frac{1}{n} + \frac{(x_0 - \bar{x})^2}{\Sigma(x_i - \bar{x})^2}\right]$ — the variance of the predicted value of y given x_0

(m) $T = \dfrac{y_0 - (A + Bx_0)}{\left[S_u^2 \left[\frac{1}{n} + \frac{(x_0 - \bar{x})^2}{\Sigma(x_i - \bar{x})^2}\right]\right]^{1/2}}$ — t-statistic of the predicted value

2. This exercise introduces you to some novel ideas about constructing some matrices that you will find useful in the following chapters. Let `I←ι10`.
   (a) `I∘.=I` produces the identity matrix
   (b) `0 3[1+I∘.≤I]` produces an upper triangular matrix
   (c) `0 3[1+I∘.=I]` produces a diagonal matrix
   (d) `2 3[1+(I∘.=I)≠1]` a symmetric matrix
   (e) `10 10 ρι11` a circular matrix
   (f) `10 10 ρ2 3,(8 ρ0),3` a tridiagonal matrix
   (g) `V∘.=V` where `V←1 1 1 2 2 2 2 3 3 3 3 3` , a block diagonal matrix

3. The following exercises are basic to calculations in *ANOVA* problems.
   For any matrix $X_{n \times k}$ where n is the number of rows and k is the number of columns, use your APL to write a routine to calculate each of the following:
   (a) Row means
   (b) Column means
   (c) Total mean and compare to (d) and (e)
   (d) Mean of row means
   (e) Mean of column means
   (f) The mean of the squared differences of each element from the total mean
   (g) The mean of the squared differences of each element from its own row mean
   (h) The mean of the squared differences of each element from its own column mean
   (i) The mean of the squared differences of each element from its own row maximum value

4. You might want to see where the total variation of any contingency table might come from. Let the matrices A, B, C, D represent four different sets of tabulated data of ninety observations which you wish to analyze.

$$A = \begin{bmatrix} 10 & 10 & 10 \\ 10 & 10 & 10 \\ 10 & 10 & 10 \end{bmatrix} \qquad B = \begin{bmatrix} 10 & 10 & 10 \\ 4 & 10 & 16 \\ 16 & 10 & 4 \end{bmatrix}$$

$$C = \begin{bmatrix} 8 & 8 & 14 \\ 4 & 14 & 12 \\ 9 & 8 & 13 \end{bmatrix} \qquad D = \begin{bmatrix} 8 & 8 & 8 \\ 4 & 14 & 12 \\ 15 & 8 & 13 \end{bmatrix}$$

Use your program in Exercise 3 to get all the quantities given by the program for all four matrices. Comment on the results.

★ **5.** Let $F(P) = \int_0^\infty x^{P-1}e^{-x}dx$

Find $F(20)$ and $F(3.5)$ using the generalized APL factorial.

Statistical Applications

1. You are given the following data:

X	65	63	67	64	64	68	62	70	66	68	67	69	71
Y	68	66	68	65	69	66	68	65	71	67	68	70	65

where X is the height of the father and Y is the height of the son, both measurements taken to the nearest inch. The objective is to find out if the height of the son depends on the father's height, and if so what is the specific relationship between the fathers' and sons' heights. Use your *Y REGRESS X* function (page 146) to find out which one of the following regressions fits the data best.

(a) $Y = a + bx + u$
(b) $Y = a + bx^2 + u$
(c) $Y = a + b \ln x + u \quad \Leftrightarrow e^y = e^a x^b e^u$
(d) $\ln Y = \ln a + x + u \quad \Leftrightarrow Y = ae^x e^u$
(e) $\ln Y = \ln a + b \ln x + u \quad \Leftrightarrow Y = ax^b e^u$
(f) $Y = a + (b/x) + u$

Use as your criterion for best fit that regression which produces the highest sample value for the coefficient of determination.

For each equation plot the residuals $\hat{u} = Y - \hat{Y}$ against $\hat{Y}$. Comment on your observations.

2. Use the routine *Y REGRESS X* (page 146) to solve the following problem.

You are given the data:

X	64	71	53	67	55	58	77	57	56	51	40	68
Y	57	59	49	62	51	50	55	48	52	42	30	57
Z	8	10	6	11	8	7	10	9	10	6	2	9

where X = weight to the nearest pound, Y = height to the nearest inch, and Z = age to the nearest year of 12 boys.

Run the following regressions in order to discover to what extent the variables X, Y, and Z are linearly related.

a_1 : $X = a + by + u_1$
a_2 : $Y = a + bx + u_2$
a_3 : $Y = a + bz + u_3$
a_4 : $X = a + bz + u_4$

After making the required transformations, run the regressions

a_5 : $\log_{10} Y = a + b \log_{10} X + u_5$
a_6 : $Y = a + b/x + u_6$
a_7 : $Y = a + b \log X + u_7$
a_8 : $Y = a + b \log Z + u_8$
a_9 : $Z = ax^b e^{u9}$ and $\ln Z = \ln a + b \ln x + u_9$

Explain intuitively the meaning of each equation and of the estimated $\hat{a}$s and $\hat{b}$s.

3. The following data give the yields of wheat on some experimental plots of ground corresponding to four different sulphur treatments for the control of rust disease. The treatments consisted of:

(1) dusting before rain
(2) dusting after rain
(3) dusting once each week
(4) no dusting

Test to see if there are any significant differences in the yields due to the dusting methods.

	Dusting method			
Plot	1	2	3	4
1	5.3	4.4	8.4	7.4
2	3.7	5.1	6.0	4.3
3	14.3	5.4	4.9	3.5
4	6.5	12.1	9.5	3.8

4. The number of units of work done per day by five workers using four different types of machines is given in the following table. Each worker operated each type of machine for one day. Find estimates of the differential effects due to

(a) machine type
(b) worker's skill

In each case, specify carefully the maintained, null, and alternate hypotheses

Units of Work Output by Type of Machine and Worker

	Machine Type			
Worker	I	II	III	IV
1	40	40	48	36
2	40	42	50	48
3	35	37	45	32
4	42	36	48	30
5	36	40	50	40

(c) Now suppose that you were not able to get the observation (3, III), i.e., the number of units for the third worker using the third machine. How would you answer questions (a) and (b)?
Some suggestions are:
1. Put some row or column average in the (3, III) position.
2. Change the routine in such a way that you need use only three observations for the third row and four observations for the third column.

5. Put the cumulative F distribution function (page 163) into your workspace. Add some lines to ANOVA1 and ANOVA2 functions already defined in order to give the answers of hypotheses tests immediately, so that you don't have to look up the F tables.

6. Add some lines to the function `Y REGRESS X` to obtain the following statistics.
   (a) the vector of calculated Y, call it $\hat{Y}$
   (b) the vector of calculated residuals, call it $\hat{U} = Y - \hat{Y}$
   (c) the mean of X, Y, and $\hat{Y}$
   (d) var $(\hat{U})$ and var $(\hat{Y})$

7. Poultry researchers investigated the weight gains (in pounds) of four types of "super" or "Industrial" turkeys fed three different rations over a period of several months. The results are listed below:

	Type of Ration		
Type of Turkeys	1	2	3
A	50	45	35
B	41	38	45
C	61	35	55
D	55	59	61

Find estimates of the differential effects of (a) rations, and (b) types, and test the hypotheses:

(a) H_0: The variations of weight gains is due to the different rations, regardless of the type of turkey.
against H_1: The rations have no effect on gains.

(b) H_0: The variation of gains is due to the type of turkey, regardless of the rations.
against H_1: The type of turkey does not affect gains.

(c) H_0: Neither the type nor the ration affects the weight against H_1: At least one of type or ration affects the weight.

8. The following diagram will help you understand the role of the degrees of freedom (n) of the X^2 distribution. Let X^2 represent the random variable and $f(X^2)$ the corresponding density function.

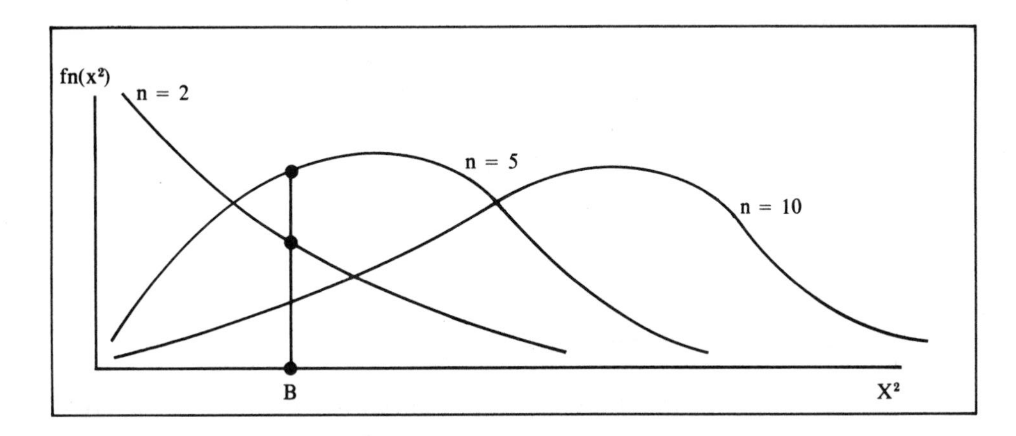

Notice that for the same upper bound (B) and increasing degrees of freedom the value of the integral, $\int_0^B f_n(x^2)dx^2$, decreases in n.

(a) Use your *N CUMCHISQ B* function (page 161) to verify this for $N =$ 1, 2, 3, . . . , 20 and $B = 5$.

(b) Use the *N CUMCHISQ B* function (page 161) to calculate the integrals for $N = 20$ and $B = 1, 2, 3, 4, 5, 6, 7, 8, 9, 10$.

(c) Alternatively. How would you use the *N CUMCHISQ B* function to determine the value of B for which the area from 0 to B is 87% when $N = 20$? Notice that the value of 87% is not in the tables of the chi-square distribution given by most textbooks.

10

Matrix Algebra in APL— How Simple It Is

10.1 Vectors, Matrices, and Arrays

Up to this point in our discussion we have dealt with variables which can be scalars or which can be arrays; arrays have one or more dimensions, scalars are dimensionless. You will recall that if we define the variable V by $V \leftarrow$ "some number," ρV produces "blank" because a scalar has no dimension, and $\rho\rho V$ (which is the dimension of the number of dimensions) produces 0. But if V is defined by $V \leftarrow$ "a list of numbers," ρV produces the number of elements in the list and $\rho\rho V$ produces 1 (the number of elements in the list of dimensions).

Vectors, Matrices, and Arrays

Chapter 9 introduced a very important extension to our definition of variables—variables defined as two-dimensional arrays which can be visualized as a table of entries with rows and columns. In this chapter we will develop our APL tools for handling two-dimensional arrays and we will distinguish between APL expressions such as "arrays" and mathematical expressions such as "vector" and "matrix." In Chapter 11 we will extend the examination of arrays in APL to three and higher dimensions!

Following the theme of this book, we will relate the APL expressions directly to the mathematical operations you have learned, or are learning, in matrix algebra. Even if you do not know any matrix algebra at all, this book will help you to learn some.

A matrix is a two-dimensional array. Its dimensions, say (n, m) (sometimes this is written as $(n \times m)$, signify that the matrix has n *rows* and m *columns*. But what if either m or n is 1? If the matrix is $1 \times m$ (one row, m columns), it is called a row vector (or often just a vector for short). And if the matrix is $n \times 1$ (n rows, 1 column), it is called a column vector (just a vector for short). So you see that when referring to a vector, you have to be careful to distinguish between row vectors and column vectors.

Arrays, Column Vector, Row Vector

Now a row vector is not a column vector is not a list! (Remember, a list has only one dimension: length.) "Array" in APL is a general term indicating a variable with one or more dimensions. An array with one dimension is a list; an array with two dimensions is a table. Mathematically, we need to be a bit more precise, so we will often have to be careful to distinguish row and column vectors from each other, from matrices, and especially from lists.

If you have a list, say *A*, and you want to make it a vector, either column or row, then use dyadic ρ. For example,

```
      A← 1 2 3 4
```

A is a list.

Reshape ρ

```
      CV←4 1 ρA
```

CV is a column vector, or a matrix of dimensions (4, 1).

```
      RV←1 4 ρA
```

RV is a row vector, or a matrix of dimensions (1, 4).

Now type

```
      A
1 2 3 4
      CV
1
2
3
4
      RV
1 2 3 4
```

A, *CV*, and *RV* are all arrays.

```
      ρA
4
```

{4 elements in a list.

```
      ρCV
4 1
```

{4 rows, 1 column in a column vector.

```
      ρRV
1 4
```

{1 row, 4 columns in a row vector.

Now try:

Ravel ,

```
      B←,CV
      ρB
4
```

{*Ravel* “,” converts the column vector *CV* into a *list*.

```
      D←,RV
      ρD
4
```

Again, the monadic function ravel converts the row vector *RV* into a list.

So now you know that lists and vectors are different, how to get vectors from lists (use reshape ρ), and how to get lists from vectors (use ravel ,).

10.2 Elementary Matrix Operations

Our first task is to define some matrices to use as examples. Try

```
      A23←2 3 ρ1 2 3 4 5 6
      A13←1 3ρ1 2 3
      A31←3 1ρ3 2 1
      A33←3 3ρ1 2 3 3 2 1 4 2 3
      B23←2 3 ρ¯1 2 3 ¯4 5 6
      B31←3 1ρ4 1 2
      B13←1 3ρ4 1 2
```

Naming of matrices is arbitrary. You could have used *DOG* or *MOUSE*, etc. Our name choice was intended to make the dimensions easier to understand.

Now that we have defined a number of two-dimensional arrays, we will need to consider how to refer to individual elements of an array. Consider

```
      A33[1;2]
2
      A23[2;3]
6
      B31[2;1]
1
```

That's easy enough, but what if we want to refer to a whole row or column of *A*33? How could we do that? Try

```
      A33[1; ]
1 2 3
      A23[ ;2]
2 5
```

Notice that with `A33[1; ]` and, more strikingly, with `A23[ ;2]` the array returned is a *list*, not a row or column vector. Thus

```
      ρA33[1; ]
3
      ρA23[ ;2]
2
```

Now try

```
      A33[1 2; 1 2]
1 2
3 2
```

{This is the first two rows and the first two columns of $A33$.

or

```
      A33[1 3; 2 1]
2 1
2 4
```

```
        (2,1)
      1→① ②  3
(1,3)   3  2  1
      3→④ ②  3
        ↑  ↑
        1  2
```

{First and third rows with second and first columns.

or again

```
      A33[1 2; 3]
3 1
```

{Gives the 3rd element in rows 1 and 2.

Matrix Addition and Subtraction

Matrix Addition and Subtraction

Matrices of the *same* dimensions can be added and subtracted.

```
      A23+B23
0 4  6
0 10 12
```

but if you try

```
      A23+A31
LENGTH ERROR
      A23+A31
      ^
```

This is because in APL, addition, subtraction, multiplication, and so on, are operations which are carried out between *corresponding* elements in the two matrices. This is nothing more than an extension to two dimensions of the idea we met before in adding, subtracting, etc., one-dimensional arrays or lists.

Multiplying a Matrix by a Scalar

Multiply Matrix by a Scalar

Multiplying a matrix by a scalar is obviously easy:

```
      3×A23
3  6  9
12 15 18
```

and forming linear combinations of vectors is also easy. Try the mathematical relation 2(*A*31) + 3(*B*31); in APL this is

```
      (2×A31)+3×B31
18
7
8
```

What would happen if we removed the parentheses? Try it.

Matrix Multiplication

The mathematical operation of *matrix multiplication* (to be distinguished from the *APL operation* of multiplication between arrays, e.g., *A*23 × *B*23) is nothing more than the APL function called inner product (defined on page 27) between the rows of the first matrix and the columns of the second. The mathematical expression for the matrix multiplication of two matrices *B*23, *A*33 to give a matrix *C* is *C* = (*B*23) (*A*33), where *C* has dimensions (2 × 3).

```
      C←B23+.×A33
      C
17 8  8
35 14 11
```

For example, the [1;2] element of *C* is the inner product of row [1;] of *B*23 and column [;2] of *A*33. Let's check that.

```
      B23[1;]+.×A33[;2]
8
```

or, to obtain *C* [2;2], try

```
      B23[2;]+.×A33[;2]
14
```

Matrix Multiplication +.×

The APL operation of +.× for obtaining the mathematical operation of inner product between vectors or, more generally, "matrix multiply," while very useful, is not without some dangers. What if by mistake one of the matrix variables is *not* defined to be a matrix, but is only a list? Consider

```
      BT←4 1 2                  {BT is a list of dimension 3.
      GT←A33+.×BT
      GT
12 16 24                        (The result of the operation +.×
      ρGT                       {between a matrix (3 × 3) and a
                                (list is a list of dimension 3.
3
```

The importance of noting the difference between `A33+.×BT` and `A33+.×B31` is illustrated by the following. Try

```
      AT←4 5 6                  {AT is a list of dimension 3.
      A13+GT                    {A13 is 1 × 3, but GT has only
                                {dimension 3.
RANK ERROR
      A13+GT
      ∧
```

because

```
      ρA13
1 3
      ρAT
3
```

But

```
      AT+GT                     {Both are lists.
16 21 30
```

and

```
      A31+(A33+.×B31)
15
18                              {A column vector (matrix which
                                {is 3 × 1).
15
```

because

```
      ρA33+.×B31
3 1
```

```
      ⍴A31
3 1
```

So the rule is: to be safe in carrying out matrix operations, *do not* mix lists with vectors *and* make sure *all* your arrays are dimensioned as matrices and vectors of the *appropriate shape*. That is, you should always know whether you are dealing with a row or a column vector. This last point is a tricky one. Consider

```
      G←A33+.×B31
      G
12
16
24
```

But what if we take the inner product between $A33$ and $B13$, which operation is not defined mathematically as a "matrix multiplication"?

```
      A33+.×B13
24 6 12
24 6 12
36 9 18
```

It is unfortunate here that you do not get an error message telling you that you cannot multiply a (3×3) matrix by a (1×3) vector. What happens in APL is that each element of $B13$ is treated in turn as a scalar, and each *row* of $A33$ is multiplied by that scalar and added up to get the above result. Thus

$$\begin{array}{lll} 24 = (1 + 2 + 3) \times 4 & 6 = (1 + 2 + 3) \times 1 & 12 = (1 + 2 + 3) \times 2 \\ 24 = (3 + 2 + 1) \times 4 & 6 = (3 + 2 + 1) \times 1 & 12 = (3 + 2 + 1) \times 2 \\ 36 = (4 + 2 + 3) \times 4 & 9 = (4 + 2 + 3) \times 1 & 18 = (4 + 2 + 3) \times 2 \end{array}$$

Compare this result with `A33+ . × B31`.

10.3 Transpose of a Matrix

Matrix Transpose ⍉

Matrices can be transposed. That is, a $(m \times n)$ matrix is converted to an $(n \times m)$ matrix simply by writing all its rows as the columns of the transposed matrix (and hence all its columns as rows in the transposed matrix). The primitive APL function (i.e., it's on the keyboard!) which does this is Transpose, ⍉ (key upper shift *O*, backspace, and key \). Thus

```
      ⍉B23
¯1 ¯4
2 5
3 6
```

Transposition of a list does nothing, since a list has length only. Try transposing *AT*.

10.4 A Not So Elementary Operation: Matrix Inverse

Matrix Inverse

Let's begin with the simplest situation—a square matrix which is nonsingular. A square matrix is one with as many rows as columns. A nonsingular matrix is one for which *no* linear combination of the rows or columns will produce a vector of zeros. More formally, a matrix A with rows $a_1, a_2, \ldots, a_n$ or columns $a^1, a^2, \ldots, a^n$ where we can find *no* list of n numbers (not *zero* of course) to satisfy either

$$a_1 b_1 + a_2 b_2 + \cdots + a_n b_n = 0$$

or

$$a^1 b_1 + a^2 b_2 + \cdots + a^n b_n = 0$$

Identity Matrix

is nonsingular. An important and very special example of a nonsingular matrix is the identity matrix, I. I looks like this:

$$\begin{matrix} 1 & 0 & 0 & 0 & 0 & . & . & . & 0 \\ 0 & 1 & 0 & 0 & 0 & . & . & . & 0 \\ 0 & 0 & 1 & 0 & 0 & . & . & . & 0 \\ . & & & & & & & & . \\ . & & & & & & & & . \\ . & & & & & & & & 0 \\ 0 & 0 & . & . & . & . & . & 0 & 1 \end{matrix}$$

That is, I is square and has zero's everywhere except for the 1's on the diagonal.

Singular and Nonsingular Matrices

The interesting thing about a nonsingular matrix is that we can always find another matrix, say B, which satisfies the following mathematical relationship:

$$AB = BA = I$$

where AB and BA represent matrix products. B is said to be the "inverse" of A. The above relationship between A, B, and I is similar to $n(n^{-1}) = 1$, where n is any nonzero number. So instead of talking about B, let's complete the analogy with numbers and write A^{-1} for "A inverse." We have now

$$AA^{-1} = A^{-1}A = I$$

and if A is a (1×1) dimensional matrix, we get the same result that we would get if A were a scalar, namely

$$aa^{-1} = a^{-1}a = I$$

You will recall that, for a a number, a^{-1} is called the reciprocal, or multiplicative inverse.

With numbers we know that ÷N gives DOMAIN ERROR if N has the value 0. The matrix analogue to $N = 0$ is a singular matrix; singular matrices do not have inverses. Consider a few simple matrices:

```
1 0       is nonsingular
0 1

1 3       is singular
2 6

1 2 3     is singular
4 5 6
7 8 9

1 2 3     is nonsingular
4 6 5
8 7 9
```

Try adding together combinations of rows (or of columns) in order to get a zero vector. More easily, see that you cannot find a combination of two of the rows (or columns) which equals the negative of the third for the nonsingular matrix.

Quad-divide ⌹ or Domino

This is all very well, but how can we get the inverse of a matrix in APL? Fortunately, we have another primitive function called "quad-divide," "domino," or "matrix-divide." It is the monadic ⌹, keyed in by upper shift L backspace, key ÷. Consider the following matrix.

```
      A←2 2 ρ5 0 0 2
      A
5 0
0 2
      ⌹A
0.2          0
0          0.5
      A+.×⌹A
1            0
0            1
      (⌹A)+.×A
1            0
0            1
```

Let's calculate the inverse of $A33$ and check that it is in fact the inverse.

```
      ⌹A33
¯3.3333E¯1    ¯1.3064E¯16    3.3333E¯1
 4.1667E¯1     7.5000E¯1    ¯6.6667E¯1
 1.6667E¯1    ¯5.0000E¯1     1.0000E0
      A33+.×⌹A33
 1.0000E0      3.1345E¯16   ¯2.6368E¯16
 5.5511E¯17    1.0000E0      9.7145E¯17
 4.4409E¯16   ¯7.8469E¯17    1.0000E0
```

As long as we recognize that numbers on the order of 10^{-16} are essentially zero, we see that we did, in fact, get the inverse of A. Try `(⌹ A33) +.×A33` on your own. From these examples you see that the results of the operation of taking the inverse may be only approximate and not exact. In any case, we see that `⌹ A33` is a very good approximation to $A33^{-1}$, the inverse of $A33$.

Domino, or quad-divide, just like ÷ for numbers, also has a dyadic use that we will discuss further in a moment. For now, note that all you needed to do to check that `(⌹ A33)+.×A33` is approximately I_3 was to type

```
      A33⌹A33
 1.0000E0     8.7093E¯17    2.7911E¯16
 2.2214E¯16   1.0000E0     ¯2.4126E¯16
¯1.3362E¯16   1.7367E¯16    1.0000E0
```

If you are wondering why you get slightly different results in the two cases, the answer is that this is due to the use of different algorithms (computational procedures) for using ⌹ dyadically and using `+.×` with monadic ⌹.

This is simple enough if you have been reading some matrix algebra, but ⌹ in APL enables you to consider something a little more mathematically tricky. We just reminded you that for a square matrix A, A^{-1} is a matrix which satisfies two mathematical conditions:

(a) $A^{-1}A = I$

(b) $AA^{-1} = I$

Left and Right Inverse of a Nonsquare Matrix

where I is the identity matrix of the same dimensions (square) as A and, of course, A^{-1}. But what if we have a nonsquare matrix, say A, which is $(m \times n)$, and there exists a matrix such that only *one* of the conditions holds? We can define a left and a right inverse, say A_{LI} and A_{RI}, depending upon which condition is satisfied. It is only when A is square and nonsingular that both left and right inverses exist and are equal (i.e., $A_{\text{LI}} = A_{\text{RI}}$) so that we can talk about the inverse of A. In this case $A^{-1} = A_{\text{LI}} = A_{\text{RI}}$. Thus if we can find a matrix A_{RI} so that

$$\underset{(m \times n)}{A} \quad +.\times \quad \underset{(n \times m)}{A_{\mathrm{RI}}} \quad = \quad \underset{(m \times m)}{I}$$

then A_{RI} is the right inverse of A. Similarly, if

$$\underset{(n \times m)}{A_{\mathrm{LI}}} \quad +.\times \quad \underset{(m \times n)}{A} \quad = \quad \underset{(n \times n)}{I}$$

then A_{LI} is said to be a left inverse (it multiplies A *from* the *left*).

Quad-divide applied to a nonsquare matrix A will yield the *left* inverse of A if that inverse exists. A left inverse of A can be found if the columns of A are linearly independent vectors, which means that no column can be set equal to a linear combination of the other columns. If the *columns* of A are linearly independent, A is said to have full column rank. Let's try it. We need a nonsquare matrix with *at least as many rows as columns*. Let's use the transpose of $A23$, which gives us 3 rows and only 2 columns. Key in

```
      ⍉A23
1 4
2 5
3 6
```

⍉A23 is of full column rank because one cannot find a number C such that

$$\begin{bmatrix} 1 \\ 2 \\ 3 \end{bmatrix} C = \begin{bmatrix} 4 \\ 5 \\ 6 \end{bmatrix}$$

Now try

```
      ⌹⍉A23
¯0.94444    ¯0.11111     0.72222
 0.44444     0.11111    ¯0.22222
```

If this is to be a left inverse, premultiplying ⍉A23 by ⌹⍉A23 should give I_2:

```
      (⌹⍉A23)+.×⍉A23
1.0000E0      1.3323E¯15
6.9389E¯17    1.0000E0
```

Here we see that ⍉⌹⍉A23 would give us the right inverse of $A23$. From this you will realize that the *right inverse* of $A23$ is the transpose of the left inverse of the transpose of $A23$. Try

```
      A23⍉⌹⍉A23
1.0000E0      6.9389E¯17
1.3323E¯15    1.0000E0
```

You will notice that the matrix result printed out is the transpose of the previous matrix result.

Systems of Linear Equations

System of Linear Equations

Now that we are equipped with quad-divide, we have an easy way to solve linear equations and to obtain linear least-squares estimates of coefficients in a much simpler way than the approach of Chapter 9. Consider first the linear equation

$$Y = XB$$

Where Y is an $(n \times 1)$ vector, X is an $(n \times m)$ matrix, and B is an $(m \times 1)$ vector. Clearly Y is obtained as the inner product (*mathematical* operation of matrix multiply) of X and B. Suppose we know X and Y but not B and, perverse creatures that we are, we want to know B as well. We now see that the solution is apparent. If X has more rows than columns $(n > m)$ and X has full column rank (rank of $X = m$), then all we need is the left inverse of X, say X_{LI}. Thus if $Y = XB$, then $X_{\mathrm{LI}}Y = X_{LI}XB = IB = B$. The way to write $X_{\mathrm{LI}}Y$ in APL is `Y⌹X` using the dyadic form of quad-divide or, more obviously but less elegantly, `(⌹X) +.× Y`.

Let's create a Y vector. Let X be the transpose of $A23$ and let B be given by `B←2 3`. Then Y is defined by $Y = X'B$ or, in APL,

```
      B←2 3
      Y←(X←⍉A23)+.×B
      Y
14 19 24
```

Remember, Y is a list, not a vector, because B was defined only as a list. If we solve for our known list B, we have

```
      Y⌹X
2 3
```

Linear Regression

Will this approach work in the least-squares analysis of linear regression? Consider the regression model we solved in Chapter 9:

$$Y = a + bX + U$$

where Y and X are observed lists, U is an unobserved list, and a and b are the coefficients to be estimated. If we have n observations on Y and X, the model can be written in matrix form as follows:

$$Y = ZB + U$$

where Y is an $(n \times 1)$ vector, Z is the $(n \times 2)$ matrix shown below, B is a (2

$\times$ 1) vector with elements (a, b), and U is an $(n \times 1)$ vector of unobserved error terms. Z is given by

$$Z = \begin{matrix} 1 & x_1 \\ 1 & x_2 \\ \cdot & \cdot \\ \cdot & \cdot \\ \cdot & \cdot \\ 1 & x_n \end{matrix}$$

If we use our current approach, how might we solve the problem? Well, Z has more rows than columns, so what do we get if we use the left inverse? When we multiply

$$Y = ZB + U$$

by Z_{LI} we get

$$Z_{\text{LI}} Y = B + Z_{\text{LI}} U$$

and if we feel justified in "ignoring" the term $Z_{\text{LI}}U$, we have solved our problem: our estimate of B is $Z_{\text{LI}}Y$. But what of the "least-squares" solution, and how does it compare to our dyadic quad-divide? The least-squares approach to the problem is to define the estimator $\hat{B}$ of B by

$$\hat{B} = (Z'Z)^{-1}Z'Y$$

where the symbol Z' means the transpose of Z, $Z'Z$ is a square matrix, and $(Z'Z)^{-1}$ is its inverse (both left and right). Before worrying about the statistical meaning of this, consider the term $(Z'Z)^{-1}Z'ZB$. $(Z'Z)^{-1}$ is the inverse of $Z'Z$, so $(Z'Z)^{-1}Z'ZB = B$, but this in turn means that $(Z'Z)^{-1}Z'$ is a *left inverse of Z!* We have, therefore, reached the inescapable conclusion that Y⌹Z, which produces the inner product of the left inverse of Z with the vector Y, gives us the least-squares solution. To further establish this result, let's redo our calculations from Chapter 9. First we set up the data:

```
Y←55 70 90 100 90 105 80 110 125 115 130 130

X←100 90 80 70 70 70 70 65 60 60 55 50

Z←12 2⍴1

Z[;2]←X
```

and now all we have to do is to type

```
B←Y⌹Z

B

210.44 ¯1.5778
```

which is exactly the same solution we got in Chapter 9.

Summary

Scalars: single numbers, with no dimension.

Lists: variables that have one dimension—length.

Arrays: variables that have at least one dimension.

Matrix, or table: an array with two dimensions (rows and columns).

Vector: a special case of a matrix—a *column* vector is a matrix with *one column,* a *row* vector is a matrix with *one row*.

Reshape, dyadic use of ⍴: reshapes an array according to the specification of the left argument; e.g., *I*⍴ *A* where *A* is an array and *I* is a list of integers reshapes *A* according to the dimensions specified in *I*.

Ravel, monadic use of ",": converts the argument (array, list, or scalar) into a *list*.

Indexing of arrays: *A*[*A*;*B*;*C*;] indicates that the *a*th plane of the *b*th row of the *c*th column is being referenced; or *A*[*A*; ;*C*;] refers to the *a*th plane and *c*th column for *all* rows.

Arrays and arithmetic functions: *A f B* produces an array *C*, each of whose elements are defined by the functional relationship *f* between the *corresponding* elements of *A and B*. *A* and *B* must have the same dimensions.

Matrix multiplication: obtained in APL by use of the inner product function +.×; e.g.,*A* +.× *B*, where *A* and *B* are matrices, produces an arracy *C* whose elements are given by the mathematical operation of matrix multiplication.

Transpose, ⍉: keyed by upper shift *O*, backspace, \. In its monadic use it alters an array so that its dimensions are reversed. If ⍴ *A* is 3 5 2, ⍴⍉*A* is 2 5 3.

Matrix transposition, *A*' (or A^T): given by ⍉*A*.

Matrix inverse and identity matrix: The identity matrix, *I*, is a square matrix with 1's on the diagonal and zero's elsewhere. A square matrix *A* has an inverse A^{-1} if $AA^{-1} = A^{-1}A = I$, where $A^{-1}A$ represents matrix multiplication. A matrix is said to be nonsingular if its inverse exists.

Quad-divide (domino), ⌹: keyed by upper shift *L*, backspace, ÷ (monadic use). If *A* is an array of *two* dimensions with the first at least as large as the second, then ⌹ *A* produces the left inverse of *A*, A_{LI}; e.g., A_{LI} +.×*A*=*I*, where the dimensions of *I* are the same as the *second* dimension of A.

Quad-divide (dyadic use): Solves linear equations. For example, if you wish to find the array *X* such that $AX = B$, where *B* is a list and *A* is a matrix, then *X* is given by *B*⌹*A*. The result *B*⌹*A* is equivalent to (⌹*A*)+.×*B*. In a linear regression $Y = XB + U$, the estimator for *B* can be obtained by *Y*⌹*X*, which produces a result equivalent to the mathematical expression $(X'X)^{-1}X'Y$.

Exercises

APL Practice

1. Let `A←4 4 ρι5`
   (a) Write the APL statements to select the first column of A?
   (b) Write the APL statements to select the second row of A?
   (c) Write the APL statements to replace all the elements that are equal to 5 with the number 6?
   (d) Write the APL statements to instruct the computer to multiply the first row by the last row element by element?
   (e) Write the APL statements to express each element of the matrix as a percentage of the largest element of the matrix?
   (f) Suppose you want each of the 4 column averages and each of the 4 row averages. Write a routine to do this.
   (g) Write the APL statements to select the two by two middle block of the matrix A?

2. Let `A←2 2ρ 26 16 9 6`, `B←2 2 ρ3 5 1 2`, `C←26 9` , and `D←1 2 ρ16 6`. Examine carefully the results of:

   (a) `⌹ B`
   (b) `B ⌹`
   (c) `⌹ ⍉ B`
   (d) `⍉ ⌹ B` and compare to (c)
   (e) `D ⌹ C`
   (f) `A ⌹ C`
   (g) `A ⌹ B` ·
   (h) `(,D) ⌹ B`
   (i) `C ⌹ B`
   (j) `( ⌹ A)+.×A`
   (k) `A ⌹ ⌹ B`
   (l) `A+.× ⌹ A`
   (m) `C ⌹ ⌹ ⌹ B`
   (n) `A ⌹ A`
   (o) `2 ⌹ 2`

3. Solve the following systems of equations.
   (a) $3X_1 + 5X_2 = 26$
   $X_1 + 2X_2 = 9$
   Compare this solution with the one for 2(i) above.
   (b) $3X_1 + 5X_2 = 16$
   $X_1 + 2X_2 = 6$
   Compare this solution with the one for 2(h) above.
   (c) Compare the results of both (a) and (b) with those found in 2(g) above.

4. Find the left inverse of the matrix `Z←3 2 ρ3 5 1 2 4 5` and verify that `((⌹Z)+.×Z) = (⍉Z)+.×⍉⌹Z` .

5. Consider the matrix `W←3 2 ρ1 2 2 4 3 6`,which does not have full column rank. Use APL to verify that W does not have a left inverse.

6. Consider the following system of equations:
   $6X_1 + 4X_2 + 3X_3 = b_1$

$20X_1 + 15X_2 + 12X_3 = b_2$
$15X_1 + 12X_2 + 10X_3 = b_3$

(a) Find the values of X_1, X_2, and X_3 if $b_1 = 13.1$, $b_2 = 46.9$, and $b_3 = 37.1$.

(b) Suppose you round the values of b_1, b_2, and b_3 to the nearest integer, i.e., $b_1 = 13$, $b_2 = 47$, and $b_3 = 37$. Solve the system using the rounded values for the vector of b's. Compare the solution values obtained to your answer to (a). This exercise is an example of ill-conditioning; i.e., small changes in the values of the numbers in the problem lead to large changes in the solution values.

7. Consider the following system of equations.

$5X_1 + 3X_2 = 7$
$2X_1 + 3X_2 = 5$
$3X_1 + 4X_2 = 7$

Let:

$$B = \begin{bmatrix} 5 & 3 \\ 2 & 3 \\ 3 & 4 \end{bmatrix} \text{ and } C = \begin{bmatrix} 7 \\ 5 \\ 7 \end{bmatrix}$$

(a) Compare the solutions to the following operations.

1. (⌹B)+.×C
2. C ⌹ B
3. (⍉C)+.×⍉⌹B

Comment on the results.

(b) Try the following operations:

1. ⌹ B The left inverse of B.
2. ⌹ ⍉ B The inverse of the transpose of B does not exist.
3. ⍉⌹⍉B The same as in 2.
4. ⍉ ⌹ B The transpose of the inverse.

Thus for any matrix $A_{n \times k}$ with $n < k$ the left inverse does not exist and the right inverse is ⍉⌹⍉ A while if $n > k$ the right inverse does not exist and the left inverse is ⌹ A.

8. Using the logical function equal (=) and the matrices given in exercise 2, verify the following properties.

Mathematically	APL	Explanation
(a) B=(B')'	B=⍉⍉ B	The transpose of the transpose is equal to the original matrix.
(b) (AB)'=B'A'	(⍉A+.×B)=(⍉B)+.×⍉A	The transpose of a product is equal to the product of the transposes in reverse order.
(c) (A+B)=B+A	(A+B)=B+A	Associative Law.

Mathematically	APL	Explanation
(d) $3(A+B)=3A+3B$	`(3×(A+B))=(3×A)+3×B`	Distributive Law.
(e) $(B')^{-1}=(B^{-1})'$	`( ⌹⍉ B)=⍉⌹ B`	The inverse of the transpose is equal to the transpose of the inverse.
(f) $(B^{-1})^{-1}= B$	`B =⌹⌹ B`	The inverse of the inverse is equal to the original matrix.
(g) $(AB)^{-1}=B^{-1}A^{-1}$	`(⌹A+.×B)=(⌹ B)+.×⌹A`	The inverse of a product is equal to the product of the inverses in reverse order.

9. It is often the case that you need to check if the $\lim_{n\to\infty} A^n$, where A is any square matrix, exists. In APL everything is easy. Consider the very simple function.

```
    ∇CONV A
[1]  A←A+.×B←A
[2]  →1×ι(+/+/A=B)≠×/ρA ∇
```

(a) Use the program to verify that the limits of the matrices XX and ZZ, when raised to the n power, are the Zero Matrix.

$$XX = \begin{bmatrix} 0 & 1 & 0 \\ 0 & 0 & 1 \\ 0 & 0 & 0 \end{bmatrix} \qquad ZZ = \begin{bmatrix} 0.1 & 0.2 & 0.3 \\ 0.3 & 0.1 & 0.4 \\ 0.3 & 0.2 & 0.5 \end{bmatrix}$$

(Hint: Use the trace operator to see if the matrix in fact converges.)

(b) How would you change the routine so that it would stop and print a message in case the matrix does not converge?

(c) Can we use the program to check if a matrix is idempotent? That is, if $A^2 = A'A = A$ then A is idempotent.

(d) Change the program so that it will calculate the inverse of $I - A$ using the formula $(1 - A)^{-1} = I + A + A^2 + A^3 + A^4 + \cdots$ in case A's limit is the zero matrix.

10. Let

$$X = \begin{bmatrix} 1 & 1 \\ 1 & 2 \\ 1 & 3 \end{bmatrix}$$

Find $A = I - X(X'X)^{-1}X'$, where I is the identity matrix, and verify that A is an idempotent matrix (i.e., $A^2 = A$).

11. Using

$$X = \begin{bmatrix} 1 & 1 \\ 1 & 2 \\ 1 & 3 \end{bmatrix}$$

and

$$R = \begin{bmatrix} 1 & 2 \\ 2 & 4 \\ 3 & 6 \end{bmatrix}$$

find:

(a) $(X'X)^{-1}$

(b) $(R'R)^{-1}$

(c) $[R(X'X)^{-1}R']^{-1}$

12. In the following input-output model let a_{ij} be the input of product i per unit-volume of output of product j $0 < a_{ij} < 1$, X_i be the total output of product i, and let C_i be the final demand for product i. Suppose we have only two products X_1 and X_2. The input-output equations are

$$a_{11}X_1 + a_{12}X_2 + C_1 = X_1$$
$$a_{21}X_1 + a_{22}X_2 + C_2 = X_2$$

Given the matrix of input-output coefficients

$$A = \begin{bmatrix} 0.3 & 0.1 \\ 0.4 & 0.2 \end{bmatrix}$$

find the total production of X_1 and X_2 that will meet a final demand of $C_1 = 20$ and $C_2 = 30$.

Statistical Applications

1. Given the one equation model where the coefficient of X is known

$$Y_t = 3X_t + U_t \qquad \text{and } X = \begin{bmatrix} 1 \\ 2 \end{bmatrix}$$

Consider the efficiency of the following two estimators of the coefficient of X.

$$a_1 = \frac{\Sigma_1^2 Y_t}{\Sigma_1^2 X_t} \qquad \text{and} \qquad a_2 = \frac{\Sigma_1^2 Y_t X_t}{\Sigma_1^2 X_t^2}$$

Prove that:

a. Var $(a_1) <$ Var (a_2), i.e., a_2 is more efficient than a_1 if the U_t's are independently distributed.

$$\text{and} \quad U_t = \begin{bmatrix} 1 \text{ with probability } .5 \\ -1 \text{ with probability } .5 \end{bmatrix} \quad t = 1, 2$$

b. Var $(a_2) <$ Var (a_1), i.e., a_1 is more efficient than a_2 if the U_t's are not independently distributed, but have the following joint discrete probability distribution.

U_1 U_2	Probability
(1,1)	1/10
(1,−1)	4/10
(−1,1)	4/10
(−1,−1)	1/10

IMPORTANT: In appendix E you will find two sets of data named *MACRO* and *WATT*, as well as a detailed explanation of the symbols that will be used. You are advised to store these data sets into your file because they will be used for most of the exercises in this and the following chapters.

2. Let $C = a + bY$ be the familiar Keynesian consumption function where C is the total consumption expenditures and Y is the GNP, both given in appendix E. Use the data from 1950 to 1978 to estimate the marginal propensity to consume, b, and the average propensity to consume. Comment on the relationship between them.

3. Another relationship that you probably learned in your macroeconomics courses is that imports (IM) are positively related to GNP. Use the data from the data set MACRO (appendix E) to see which of the following equations better describes the relation between imports and GNP.

(a) IM $= a + b$ GNP

(b) IM $= a + b$ GNP^2

11

Higher-Order Arrays

So far we have avoided large arrays of data and complicated statistical problems in an effort to learn the basic and easy procedures first. However, we are now getting to the stage where we can branch out and be more adventurous. You will soon see that as the complexities of the statistical problems increase, and as the amount and variety of data increase, we will need to develop new mathematical tools and hence new APL procedures to handle them.

The mathematical tool which is most heavily used in statistics is matrix algebra; it was introduced in the last chapter. What we need to do now is to learn to exploit the power of APL in solving a variety of data-handling problems and complex statistical procedures. It is to these issues that we now address ourselves.

11.1 Reduction Function

Reduction / Arrays and Indexing

In Chapter 9, when we were calculating the cell frequencies for a contingency table, we saw the need for the use of reduction across both rows and columns of a matrix. In fact, we can make the reduction function even more useful. For example, we might have a list of tables and want to get the average value of the (i, j) entries across all the tables. In APL we have a direct way of doing this, as we shall now see.

Recall from Chapter 9 that the reduction function, /, operates on only one dimension of a multidimensional array at a time. For example, the use of +/ on a matrix X of dimension 2×3 is

```
      X
1 2 3
4 5 6
```

```
      +/X
6 15
```

The summation has proceeded across the *last* coordinate of the matrix. That is, each *row* is reduced by its columns. Another way of indicating the coordinate over which reduction is taken is to specify the coordinate explicitly. For example

```
      (+/[2]X)
6 15
```

and

```
      (+/[1]X)
5 7 9
```

In the latter case, the reduction is over the first coordinate; i.e., each *column* is reduced along its rows. In this case there is an alternative procedure. By overstriking the reduction operator (/) with the subtraction symbol, producing ⌿, we can also obtain reduction over the first coordinate.

But when we have three dimensions, as in

```
      X←2 2 2 ⍴X

      X
1 2

3 4


5 6

1 2
```

where X is composed of two planes of two rows by two columns each, the situation is more complicated. In our example with X we have

	Columns 1	Columns 2	
Plane 1	1	2	← Rows 1, 2
	3	4	
Plane 2	5	6	← Rows 1, 2
	1	2	

The number "6" is in the second column of the first row of the second plane.

To find sums across the third coordinate, we could use

```
      (+/X)
3  7

11 3
```

```
      (+/[3]X)
3  7
11 3
```

which results in four sums—addition over each row in each plane. 3 is the reduction of the first row in the first plane, 7 is the reduction of the second row in the first plane, and so on.

However, the main advantage of the use of the index notation with reduction is not merely the provision of an alternative to +/ (reduction over the *last* coordinate) or to +⌿ (reduction over the *first* coordinate), but that it enables us to get reduction over the *middle* coordinate. Suppose that we want column sums in each plane. We obtain this by

```
      +/[2]X
4 6
6 8
```

Reduction over the first coordinate is obtained by

```
      +/[1]X
6 8
4 6
```

or by

```
      +⌿X
6 8
4 6
```

which gives us the sums *across* matrices of the (*i*, *j*)th elements in each matrix.

Any dyadic element can be used with the reduction operation. However, with − and ÷ you have to consider carefully what the results will be. For example, (-/1 2 3 4) will produce 1 -2 -3 -4, and in APL the value of that expression is ¯2. Also (÷/2 4 6 8) becomes 2 ÷ 4 ÷ 6 ÷ 8, which is equal to (2 × 6) ÷ (4 × 8) in APL. So you need to be concerned about both the coordinate over which the reduction takes place and the meaning of the reduction itself.

Expand \ with *Arrays*

Recall the scan instruction, \, which is similar to the reduction instruction. Scan has the same general form as reduction, and the same rules about index coordinates apply. In "sum-scan" we would have

```
      +\1 2 3 4
```

which would result in

```
1 3 6 10
```

The general form of scan is *FN**VAR*, where *FN* is a primitive dyadic function and *VAR* is a vector or matrix. The dyadic function *FN* is placed between successive *additional* elements of *VAR*. For example,

```
+\1 2 3 4
```

would be

```
1  (1+2)  (1+2+3)  (1+2+3+4)
```

The last element is exactly the same as that which you would obtain with the "sum-reduction," `+/1 2 3 4`.

Let's consider a practical example of the use of scan with three-dimensional arrays. Suppose that you had a three-dimensional array of oil production data. The planes represent different nations, the rows are oil pumping locations, and the columns are quarters of the year. We wish to know the cumulative totals by quarter for each site and each nation. For expository purposes, let us reduce the number and size of the matrices and use hypothetical values. We can construct the data by use of the roll function, `?`, which generates random numbers (see Chapter 4).

```
PRODUCTION←3 5 4 ρ60?60
```

We will pretend that the above represents the quarterly output of three nations at five sites in each nation for four quarters. In order to generate the cumulative sums by quarter, we use +/ on the array `PRODUCTION`, since quarters are the elements in the last dimension:

```
      PRODUCTION

53 15 32 27
44 60 18 17
 4 58 25  9
47 45 22 59
56 57 40 24

13 55  3  6
54 51 12 20
10 38  8 45
43 30 36 35
37  1 14 16

23 41  5 19
48 39  2 50
 7 31 21 34
42 26 28 11
33 52 29 49
```

In schematic form these data are:

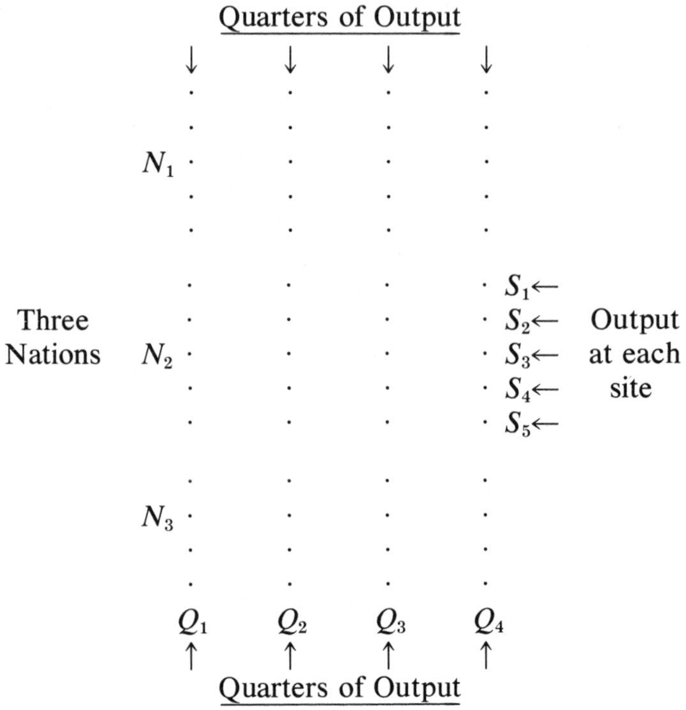

Quarters of Output

	Q_1	Q_2	Q_3	Q_4	
N_1	·	·	·	·	
Three Nations N_2	·	·	·	· S_1← … S_5←	Output at each site
N_3	·	·	·	·	

Quarters of Output

```
      +\PRODUCTION
53  68 100 127
44 104 122 139
 4  62  87  96
47  92 114 173
56 113 153 177

13  68  71  77
54 104 117 137
10  48  56 102
43  73 109 144
37  38  52  68

23  64  69  88
48  87  89 139
 7  38  59  93
42  68  96 107
33  85 114 163
```

A matrix for each nation.
Each row represents a site.
Each *column* contains the accumulating partial sums of quarterly output.

Let's look at the first nation and first site. We have 53 units of output for the first quarter, (53 + 15 = 68) units in the second, (68 + 32 = 100) in the third, and (100 + 27 = 127) in the fourth. Each of these rows yields our quarter-to-date totals, and each matrix represents the cumulative quantity output by sites for each country.

Now, if we want the total quarter-to-date figures for each country (i.e., if we want to add up over sites for each country), we type in

```
      +\(+/[2]PRODUCTION)
204 439 576 712
157 332 405 528
153 342 427 590
```

← Cumulative quarterly outputs for nation 1.
← Cumulative quarterly outputs for nation 2.
← Cumulative quarterly outputs for nation 3.

Reading from the right, we have summed down the columns of each plane and then performed "sum-scan" on each of the three vectors created by the reduction down each column.

Suppose that the second dimension represented classifications by types of oil rather than locations, and we wanted to compute the quarter-to-date sum for each of these five types *across nations*.

```
      +\(+/[1]PRODUCTION)
 89 200 240 292

146 296 328 415

 21 148 202 291

132 233 319 424

126 236 319 408
```

`+/[1] PRODUCTION` gives the totals *across* nations by type of oil and by quarter. `+\` applied to the resulting matrix gives the cumulative sums. Thus 89 = (53 + 13 + 23), and 200 = (89) + (111) = (5 + 55 + 41). In order to give you a better appreciation of the flexibility of these operations, consider the following alternatives, which are all equivalent:

```
A←+⍀(+/[1]PRODUCTION)

B←+\[1](+/[1]PRODUCTION)

C←+\[1](+⌿PRODUCTION)

D←+⍀(+⌿PRODUCTION)
```

The equivalence of these four alternatives is verified by

```
      A=B
1 1 1 1
1 1 1 1
1 1 1 1
1 1 1 1
1 1 1 1
      C=D
1 1 1 1
1 1 1 1
1 1 1 1
1 1 1 1
1 1 1 1
      A=D
1 1 1 1
1 1 1 1
1 1 1 1
1 1 1 1
1 1 1 1
```

A problem which statisticians frequently encounter is that of selecting subsets of observations. Suppose that you have a list of tables of statistics; for example, you are looking at annual observations of GNP (gross national product) statistics for a series of countries. Now while the complete list may contain as much information of this type as you would ever wish to use, and so is potentially a very useful data source, in any particular problem you may want to look at only a few variables, or a few observations, or only a selection of countries. Let's see how we can select variables from an array of several dimensions.

11.2 Compression

Dyadic Use of Compression

The compression function is the dyadic use of the reduction operator /. This operation requires a vector of 1's or zeros to the left of the "slash." Here is an example.

```
      1 0 1 0/1 2 5 7
1 5
```

What does it do? It "compresses" the vector 1 2 5 7 to 1 5, by dropping those elements from the right array which are matched by zeros on the left side of the compression symbol. This means that we must have the same number of elements on both sides of the / symbol. If we violate this rule, we get a length error.

```
      1 0 1/1 3 5 7
LENGTH ERROR
      1 0 1 / 1 3 5 7
          ∧
```

If X is a three-dimensional vector defined by

```
      X←2 2 2 ρι8
      X
1 2
3 4

5 6
7 8
```

we can compress this matrix by using the same rules we developed for scan and reduction. Here are a few examples:

```
      0 1/X
2
4

6
8
```

{Compression according to the last coordinate, columns.

```
      0 1/[3]X
2
4
6
8
```

An alternative way of doing the same thing.

```
      0 1⌿X
5 6
7 8
```

Compression according to the *first* coordinate, planes.

```
      0 1/[1]X
5 6
7 8
```

An alternative way of doing the same thing.

```
      0 1/[2]X
3 4
7 8
```

Compression according to rows, that is, according to the *second* coordinate.

Let's consider an example. Suppose that we have annual observations on GNP statistics listed as columns in a two-dimensional table, one table for each country. This mammoth variable is called *GNPSTAT*. You want to run a regression between, say, consumption and income, which are in columns 5 and 26 for countries for which the index numbers are 8, 9, 22, and 38. Also, you have decided to delete the war years from your analysis; these years have index numbers 40 to 45. Let's store our subset of data into *CYTRFAL*. This could be achieved by the procedure outlined below.

We need three arrays to compress *GNPSTAT*. Let's label them *CO* for the country compression, *VAR* for the variable compression, and *WAR* for the war years. Suppose that there are *N* countries, *M* variables, and *T* years altogether. Then we type in

```
      VAR←M ρ0
      VAR[5 26]← 1 1
      CO←N ρ0
      CO[8 9 22 38]← 1 1 1 1
      WAR←T ρ1
      WAR[40+0,ι5]← 6 ρ0
```

and now we can obtain the variable we want by

```
      REGVAR←CO/[1]WAR/[2]VAR/[3]GNPSTAT
```

As a numerical example of this, consider the variable *X* above and suppose that $M = N = T = 2$. We want the first "country," the second year, and the first variable; in this simple example that produces the number 3. Type in

```
        VAR← 1 0
        CO← 1 0
        WAR← 0 1
        REGVAR←CO/[1]WAR/[2]VAR/[3]X
3
```

11.3 Expand Function

Dyadic Use of Expand \

Deletion of variables naturally has its complement in the insertion of variables. Suppose for example that we have a matrix of data (i.e., a table of variable values) and we decide that we want to add some more variables to the array—not just tacked on as it were, but added into a specific place in the array. The way in which this can be done is by use of the dyadic expand function, \.

The expand function is analogous to the compress function. It has the same general form. It requires an array of zeros and 1's to the left of the expand operator (\). But with expand, the *number* of 1's in the *left* vector must be equal to the number of elements in the *coordinate* of the array to be expanded. Whenever a zero occurs in the *left* array, a zero is placed in sequence in the expanded array between the elements of the array on the right. Here is an example:

```
        X
1 2
3 4

5 6
7 8
        1 0 1\X
1 0 2
3 0 4

5 0 6
7 0 8
```

Expansion by columns (last coordinate); 0's are placed *between* columns as indicated by the array 1 0 1.

```
        1 0 1⍀X
1 2
3 4
```

Expansion by planes (first coordinate).

```
0 0
0 0

5 6
7 8
      1 0 1\[2]X
1 2
0 0
3 4

5 6
0 0
7 8
```

{ Expansion by rows (second coordinate).

As you can see, a zero in the left argument inserts a coordinate of zeros in the expanded array.

If we follow up the *GNPSTAT* example in §11.3, we can see that if we wanted to insert GNP statistics for a whole country we would first make room for them by

```
      1 1 1 . . . 1 0 1 . . . 1\[1]GNPSTAT
```

We would make room for a new variable for all countries and years by

```
      1 1 1 . . . 1 0 1 . . . 1\[3]GNPSTAT
```

In order to fill in years of observations originally left out of the data set we would enter

```
      1 1 1 . . . 1 0 0 0 1 . . . 1 0 0\[2]GNPSTAT
```

In this last example we have made room for two rows of data to be added at the bottom of each table; in short, we have made room for two further years of observations when we manage to get them.

Frequently, when we acquire a data set, the order of the variables or the arrangement of the tables of entries may not be convenient for our purposes. Consequently, it is often most useful to be able to reorder higher dimensional arrays—to rearrange them into a more suitable format. The next couple of functions help us to do just that.

11.4 Reverse or Rotate Function

Rotate ⌽

In its monadic form, ⌽1 2 3 4 produces 4 3 2 1 . The character ⌽ is produced by overstriking the circle (upper shift alphabetic *O*) and the residue | (upper shift *M*). If the vector on the right of ⌽ is a multidimensional array, the function obeys the same rules that we have discussed concerning the expand, compress, scan, and reduce operators. The following examples will illustrate the use of ⌽.

```
      X
1 2
3 4

5 6
7 8
      ⌽X
2 1
4 3

6 5
8 7
```

Reverses the order of *columns*, the *last* coordinate.

↑was column 1
↑was column 2

```
      ⌽[3]X
2 1
4 3

6 5
8 7
```

An alternative way of doing the same thing.

```
      ⌽[2]X
3 4 ←was Row 2, Plane 1
1 2 ←was Row 1, Plane 1

7 8 ←was Row 2, Plane 2
5 6 ←was Row 1, Plane 2
```

Reverses the order of *rows*, the *second* coordinate. Notice that row reversal is *within* planes, not across planes.

```
      ⌽[1]X
5 6 ←was Plane 2
7 8

1 2 ←was Plane 1
3 4
```

Reverses the order of *planes*, the *first* coordinate.

```
      ⊖X
5 6
7 8

1 2
3 4
```

An alternative way of doing the same thing.

In this last example, the overstriking of the circle with the subtraction sign instead of the residue sign indicates that the operation is to be over the first coordinate instead of the last.

The two-argument or dyadic form of rotate allows you to specify the "amount of rotation." By this we mean that you can reposition the elements of the array by specifying the rotation in the array to the right of ϕ. As an example, we rotate 4 2 1 6 two positions to the left by

```
      2ϕ4 2 1 6
1 6 4 2
```

Symbolically we have:
1st rotation step 2 1 6 4
2nd rotation step 1 6 4 2
or
4 2 1 6 Produces 1 6 4 2

We can rotate in the other direction by

```
      ¯3ϕ4 2 1 6
2 1 6 4
```

Here we have rotation to the *right* by three spaces. Here is how that worked:

```
      ¯1ϕ4 2 1 6
6 4 2 1
      ¯2ϕ4 2 1 6
1 6 4 2
      ¯3ϕ4 2 1 6
2 1 6 4
```

Perhaps an easier way to see these operations is to consider putting the numbers down in a circle with positions marked, and then to rotate the inner "dial" of numbers. Thus 2ϕ4 2 1 6 means rotate two positions to the left, and if you rotate the inner dial two positions to the left (counterclockwise) you will get 1 in position 1, 6 in position 2, etc.

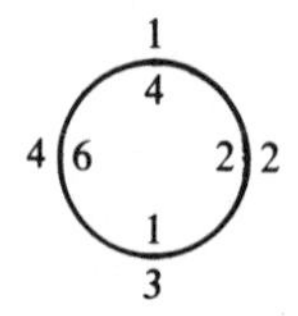

If you find the clock approach handy, use it.

In short, rotate merely *rotates* the array of numbers *in sequence,* to the right with a negative argument, and to the left with a positive argument.

This procedure is also called a cyclic shift, because although elements are rotated, they are *not* interchanged.

Just as a reminder that ¯1 and −1 are not the same, you might try

```
      -1⌽4 2 1 6
¯2 ¯1 ¯6 ¯4
```

Multidimensional arrays can be rotated along a specified coordinate. Or, you can indicate the "amount" of rotation for each row, column, or plane within a coordinate. For our example, we will use the three-dimensional array Z generated by

```
      Z←2 3 4 ρ ι24
      Z                       Row numbers
1  2  3  4                        (1)
5  6  7  8                        (2)
9 10 11 12                        (3)

13  14  15  16                    (1)
17  18  19  20                    (2)
21  22  23  24                    (3)
(1) (2) (3) (4)
```

Column numbers

```
      2⌽Z
 3   4   1   2
 7   8   5   6
11  12   9  10

15  16  13  14
19  20  17  18
23  24  21  22
(3) (4) (1) (2)
```

Rotate *columns* (last coordinate) to the *left* by two positions.

Column numbers

Next, we rotate the second dimension (rows) one place to the "right" which, for rows, is *down,* and notice that the rotation of rows is *within* planes. (Rotation of columns is also within planes, but the distinction is not apparent.)

```
      ¯1⌽[2]Z
                              Row numbers
 9 10 11 12                       (3)
 1  2  3  4                       (1)
 5  6  7  8                       (2)

21 22 23 24                       (3)
13 14 15 16                       (1)
17 18 19 20                       (2)
```

We can reverse the planes with any of the following commands:

```
      1⌽[1]Z
13 14 15 16
17 18 19 20             ← Plane (2)
21 22 23 24

 1  2  3  4
 5  6  7  8             ← Plane (1)
 9 10 11 12
      ¯1⌽[1]Z
13 14 15 16
17 18 19 20             ← Plane (2)
21 22 23 24

 1  2  3  4
 5  6  7  8             ← Plane (1)
 9 10 11 12
      ⊖Z
13 14 15 16
17 18 19 20             ← Plane (2)
21 22 23 24

 1  2  3  4
 5  6  7  8             ← Plane (1)
 9 10 11 12
```

You will gain much more familiarity with this function in the exercises at the end of the chapter. For now, it should be clear how you can rearrange your data into a more convenient format. In matrix algebra the operation of transposition of matrices and vectors is the most useful of all matrix operations. We now examine APL's development of this basic, but simple, notion.

11.5 Transpose Function

Transpose ⍉

The transpose function will alter the shape of your data matrix. The transpose symbol is constructed by overstriking the circle (above the alphabetic *O*) and the reduction operator \, forming ⍉. The two-dimensional case is an exact analogue to the mathematical operation of transposition. For example,

```
      Y←2 3 ρ1 1 1 2 2 2
      Y
1 1 1
2 2 2
      ρY
2 3
      ⍉Y
1 2
1 2
1 2
      ρ(⍉Y)
3 2
```

In mathematical notation ⍉Y is written as Y^T or Y'. The three-dimensional array is sometimes confusing:

```
      ⍉ Z
 1 13
 5 17
 9 21

 2 14
 6 18
10 22

 3 15
 7 19
11 23
```

```
 4 16
 8 20
12 24
```

What has been done here is best explained by considering the *shapes* of the two arrays. The shape of Z is

```
      ρZ
2 3 4
```

but the shape of ⍉Z is

```
      ρ(⍉Z)
4 3 2
```

So we see that we have here a natural extension of the idea of transpose; in APL terms, transpose yields an array whose dimensions are the reverse of the dimensions of the original matrix, i.e., ⍉Z 2 3 4 are the dimensions of ⍉Z.

The above discussion has shown you how the monadic form of transpose, ⍉, works. There is a very important extension of this idea in the dyadic use of ⍉. Remember that in the monadic use, if ρ(Z) was 2 3 4, ρ(⍉Z) is 4 3 2. But what if we want an array of dimensions 3 2 4? We obtain such a result with the dyadic transpose and we refer to each dimension number by its position. Thus in the first position for Z, the dimension number is 2, in the second it is 3, and in the third it is 4. If we type

Dyadic Transpose

```
      1 2 3⍉Z
 1  2  3  4
 5  6  7  8
 9 10 11 12

13 14 15 16
17 18 19 20
21 22 23 24
```

we get back Z itself.

Now let us consider the more interesting alternatives. If we specify 3 2 1 ⍉ Z, we will get an array of 4 planes of 3 by 2 matrices. An element in the [$I;J;K$] position in Z is put into the [$K;J;I$] position for the array (3 2 1 ⍉ Z). For example,

```
      3 2 1⍉Z
 1 13
 5 17
 9 21
```

ρ(3 2 1 ⍉ Z) is 4 3 2. Elements in the [$I;J;K$] position are transposed to the [$K;J;I$] position.

```
 2 14
 6 18
10 22

 3 15
 7 19
11 23

 4 16
 8 20
12 24
      1 3 2⍉Z
 1  5  9
 2  6 10
 3  7 11
 4  8 12

13 17 21
14 18 22
15 19 23
16 20 24
      2 1 3⍉Z
 1  2  3  4
13 14 15 16

 5  6  7  8
17 18 19 20

 9 10 11 12
21 22 23 24
```

ρ (1 3 2 ⍉ Z) is 2 4 3. Elements in the [I;J;K] position are transposed to the [I;K;J] position.

ρ (2 1 3 ⍉ Z) is 3 2 4. Elements in the [I;J;K] position are transposed to the [J;I;K] position.

One way to visualize these changes is:

Original Coordinate Values	2	3	4
Original Index Position	1	2	3
Transposed Index Position	2	1	3
Transposed Coordinate Values	3	2	4

These operations will be particularly important in certain matrix operations needed later on in the text. For now, let us note only the more obvious benefits of the transpose function and its extension to multiple-dimensioned arrays.

First of all, a very common matrix multiplication needed in statistics is given in mathematical terms by

$$X(X'X)^{-1}X'$$

where X' represents the transpose of X and X is an $N \times K$ matrix, $N > K$. The matrix product $X(X'X)^{-1}X'$ is programmed in APL by

```
R←X+.×( ⌹ ( ⍉X)+.×X)+.×⍉X
```

Let's try it with the matrix W defined in the following manner:

```
W←4 2 ρ1 0 0 1 1 1 2 3

R←W+.× ( ⌹ ( ⍉W)+.×W)+.×⍉W

R

 0.64706     ¯0.41176     0.25529     0.658824
¯0.41176      0.35294    ¯0.058824    0.23529
 0.23529     ¯0.058824    0.17647     0.29412
 0.058824     0.23529     0.29412     0.82353
```

A more interesting example is the following. Suppose that we have time series data on some variables for different countries. The data are averaged by country. But what we would like to observe are the data arranged by year, rather than by country. That is, the original data contain a set of observations on a number of variables by year for each country. We want to rearrange the data so that for *each* year, we have observations *across* countries for each variable. Let's give a simple example.

Suppose that the matrix Z we created before represents two countries' data for three years on four variables; Z has 2 planes, 3 rows, and 4 columns.

```
         Z              Years
 1   2   3   4          (1) ┐
 5   6   7   8          (2) ├ Country 1
 9  10  11  12          (3) ┘

13  14  15  16          (1) ┐
—   —   —   —           (2) ├ Country 2
—   —   —   —           (3) ┘
↑   ↑   ↑   ↑
1   2   3   4
   Variables
```

For each year we want 3 planes, a table of entries representing observations by country on each variable; that is 2 rows and 4 columns. This is achieved by

2 3 4	← *Current* dimensions
1 2 3	← Current position
3 2 4	← *Desired* dimensions
2 1 3	← and positions

So we type

```
      2 1 3⍉Z
 1  2  3  4
13 14 15 16

 5  6  7  8
17 18 19 20

 9 10 11 12
21 22 23 24
```

and get the desired result.

Another important "data manipulation" task is to "combine" matrices and vectors to form bigger matrices; for example, we might have a matrix X of dimension $N \times M$ and another matrix Z of dimension $N \times Q$, and want to form a new matrix $W = [X\ Z]$ of dimension $N \times (M + Q)$. The next set of functions enables us to do just that and much, much more.

11.6 Ravel, Catenate, Laminate

Ravel ,
Catenate ,
Laminate ,

The ravel function "," can convert a multidimensional array into a list. Consider the array *DATA*:

```
      DATA←2 3 4 ρ10+ι24
      DATA
11 12 13 14
15 16 17 18
19 20 21 22

23 24 25 26
27 28 29 30
31 32 33 34
```

If we ravel *DATA* we get:

```
      ,DATA
11 12 13 14 15 16 17 18 19 20 21 22 23 24 25 26 27 28 29 30 31 32 33 34
```

where we have *DATA* listed by *rows* in the first plane and then in the second. Essentially, the ravel operation (like all the others that we have discussed) operates *according* to columns *within* rows within planes, within blocks, etc. You can check the shape of *DATA* and of ,*DATA* by

```
      ⍴DATA
2 3 4
      ⍴(,DATA)
24
```

To compute an arithmetic average of all the elements of the *DATA* matrix, you could write

```
      (+/,DATA)÷ ⍴ ,DATA
22.5
```

Catenation ,

The ravel function is monadic. The dyadic form of the function is called catenate or laminate. The general form of the command is *X*,[*Y*]*Z*. When *Y* is an *integer,* the function is called *catenate* and when *Y* is a *fraction,* the function is called *laminate*. Let's begin with catenate when *Y* is an integer.

Catenation "joins together," as it were, two arrays along a specified coordinate, provided that the lengths of the arrays involved are the same, that is, provided that the two arrays are conformable. If this is not the case, you get a length error. The operation can best be explained by examples.

Let us define the following arrays:

```
      A              B              C              D
1  2  3          8  9          20 21 22          25
4  5  6         10 11
      ⍴A             ⍴B             ⍴C             ⍴D
2  3             2  2             3
```

Now try

```
      A,[2]B                 B,[2]A        { Catenation here augments
1  2  3  8  9            8  9  1  2  3     { the columns
4  5  6 10 11           10 11  4  5  6
└─────┘ └───┘           └───┘  └─────┘
   A      B               B       A
      ⍴A,[2]B                ⍴B,[2]A
2  5                     2  5
```

Catenation along columns means *column* dimensions are added, 3 + 2 = 5.

```
      A,B                     B,A

1  2  3  8  9           8  9  1  2  3
                                          {Same result as above
4  5  6 10 11          10 11  4  5  6
```

We notice here that, as we might have suspected, if no coordinate is specified, catenation, if conformable, proceeds along the *last* coordinate. Now try

```
      A,[1]C                 C,[1]A

 1  2  3 } A            20  21  22 } C     { Catenation in this case
                                           { augments rows
 4  5  6 }               1   2   3 }
         } C                       } A
20 21 22 }               4   5   6 }

      ρA,[1]C               ρC,[1]A      { So row dimensions are
                                         { added, 2 + 1 = 3
3  3                    3  3

      A,[1]D               A,D

 1  2  3                1  2  3  25      { A scalar is extended to
                                         { make up an array of appro-
 4  5  6                4  5  6  25      { priate dimension for cate-
                                         { nation.
25 25 25

      ρA,[1]D              ρA,D

 3  3                   2  4
```

We see here that *scalars* can be catenated to arrays and are automatically extended for this purpose.

Examples of data manipulation using the function catenate spring to mind. Imagine that you have an array of time series data on a set of variables, and that later you acquire data on a further few years of observation on the same variables. Catenate enables you to add the new data to your old array quickly and efficiently. Another useful example to keep in mind is the perennial problem of handling the constant term in a multiple regression problem. Suppose once again that you have an array X of data on variables to explain the movements in steel prices over time, but that before running your regression you want to add the constant term; i.e., you want a *column* of 1's as the first *column* of your data matrix. Now we know how to do this without pausing for further thought:

```
      X←1,[2]X
```

gives the desired data array. Try this with our simple A matrix:

```
      A←1,[2]A

      A

1 1 2 3

1 4 5 6
```

Lamination ,

Laminate is quite different from catenate in operation because a *new* coordinate is established. The form of the command is identical to that for catenation, specifically *X*,[*Y*]*Z*. But now *Y* can take on any positive *fractional* value. The idea of laminate is to create a *new* dimension. For example, suppose that you have two matrices of dimension 3 × 4 to be laminated. You might want 2 planes of 3 × 4 matrices, 3 planes of 2 × 4 matrices, or 3 planes of 4 × 2 matrices. Laminate lets you choose. Consider the two matrices *E* and *F*:

```
      E                      F

1  2  3  4            .1   .2   .3   .4

5  6  7  8            .5   .6   .7   .8

9 10 11 12            .9  1.0  1.1  1.2

      G←E,[.5]F              H←F,[.5]E

    G                      H

1  2  3  4            .1   .2   .3   .4

5  6  7  8            .5   .6   .7   .8

9 10 11 12            .9  1.0  1.1  1.2

.1 .2 .3 .4            1  2  3  4

.5 .6 .7 .8            5  6  7  8

.9 1 1.1 1.2           9 10 11 12

      ρG                     ρH

2  3  4               2  3  4

↑                     ↑
```

{*G*, *H* are each *3*-dimensional arrays; 2 planes of 3 by 4 matrices}

{The added dimension is indicated by the ↑}

In short, specifying `[.5]E` means that we want to *add* a dimension in *"front."* Actually, any decimal between 0 and 1.0 would do as well as .5. Now, what about something like `[1.6]F`? Here we want to add a dimension *between* the existing first and second dimensions. This operation will give us 3 planes of 4 × 2 matrices. Consider the examples:

```
      E,[1.6]F                 E,[2.6]F

 1     2     3     4          1         .1
.1    .2    .3    .4          2         .2

 5     6     7     8          3         .3
.5    .6    .7    .8          4         .4

 9    10    11    12          5         .5
.9   1.0   1.1   1.2          6         .6
```

```
      ρE,[1.6]F             7          .7
3     2   4                 8          .8
      ↑                     9          .9
                           10         1
                           11         1.1
                           12         1.2
                                   ρE,[2.6]F
                          3   4    2
                              ↑
```

{The added dimension is indicated by the ↑

These examples show us how we can fit two data matrices together in a variety of ways. For example, the first case of creating 2 planes of 3 by 4 matrices might be useful in setting up an array of time series data on a group of variables by country. Alternatively, E and F might represent, respectively, the labor and capital inputs by year for various production plants. In the `E,[1.6]F` case, each of the three planes created contains each year's *labor* and *capital* inputs to the various plants. In the `E,[2.6]F` case, each of the three planes contains the transpose of these data matrices.

Let's complete this chapter by examining the ways in which we can do the opposite to the above operations; that is, we have just been discussing fitting matrices together into *higher-dimensional* arrays, now let us see how to "undo" that process.

The functions of "drop" and "take" enable us to select in a straightforward manner from a multidimensional array those elements that we wish to use.

11.7 Take and Drop Functions

Take ↑ and Drop ↓ Functions

It is possible to extract or delete some of the elements of a multidimensional array by using the take and drop functions. These functions, like the others we have described, will operate on character matrices as well as on numeric ones.

The take function, ↑, which is upper case `Y`, extracts elements of an array. It has the format `A↑F`, where A elements are extracted from B, starting on the *left*. For example

```
      2↑'ABCDE'
AB
```

but

```
      ¯3↑'ABCDE'
CDE
```

which takes three elements from the *right*.

Now consider the three-dimensional array

```
      B←2 2 3 ρ'ABCDEFGHIJKL'
      B
ABC
DEF

GHI
JKL
      1 2 2 ↑ B
AB
DE
```

takes the first plane from the first dimension, the first two rows from the second dimension, and the first two columns from the third dimension.

The drop function, ↓, removes elements from an array or list, just as the take function extracts elements. The down arrow ↓ is above the character *U* on the keyboard.

```
      3 ↓ 'ABCDEF'
```

drops the first three elements *ABC*, and we get

```
DEF
```

The code

```
      ¯3 ↓ '+/*A,?6'
```

results in

```
+/*A
```

As you can see, characters other than alphabetic or numeric ones can be operated on. The drop function also operates on multidimensional arrays, just as the take function does. For example,

```
      B
ABC
DEF

GHI
JKL
      1 0 2 ↓ B        { Drops the first plane, keeps all rows,
                       { and drops the first two columns.
I
L
      1 1 2 ↓ B        { Drops the first plane, first rows,
                       { and first two columns
```

```
L
      ¯1¯1¯1 ↓ B
AB
      1 ¯1 0 ↓ B
GHI
      0 0 0 ↓ B
ABC
DEF

GHI
JKL
```

{Drops the last plane, the last row, and the last column

{Drops the first plane, last row, and keeps all columns

{Drops nothing, i.e., keeps everything

Summary

Reduction, /, and Scan, \: across multidimensional arrays operates as with lists, but over the *last* coordinate. Overstriking / or \ with - (minus sign) yields reduction or scan over the *first* coordinate. Reduction or scan over any coordinate can be chosen by specifying the chosen coordinate by indexing. Thus if, for example, ρ*A* is 2 3 4 we have:

+/*A*	is equivalent to +/[3]*A*, reduction by columns
+⌿*A*	is equivalent to +/[1]*A*, reduction by planes
+/[2]*A*	yields reduction by rows

Similar remarks hold for the scan function.

Compression, dyadic use of /: If *A* is a list of 0's and 1's, and *B* is a list, then *A*/*B* produces a list *C* of dimension (+/*A*) which contains those elements of *B* indicated by the 1's in *A*. Compression can be used with higher-dimensional arrays in exactly the same way as can reduction or scan.

Expansion, dyadic use of \: If *A* is a list of 0's and 1's, and *B* is a list, then *A**B* produces a list *C* of dimension ρ(*A*). Then +/*A* must equal ρ*B*. *C* will contain the elements of *B* with zeros inserted in the positions corresponding to the zeros in *A*. Expansion with multiple-dimensional arrays expands the size of the indicated dimension by the number of zeros which appear. For example, if ρ*X* is 2 2 2:

1 0 1*X* is equivalent to 1 0 1\[3]*X* and produces a middle column of zeros in each matrix in each plane.

1 0 1⍀*X* is equivalent to 1 0 1\[1]*X* and produces a matrix of zeros between the two planes of matrices of *X*.

1 1 0\[2]*X* produces a *row* of zeros at the end of each matrix in each plane.

Rotate, ⌽ (keyed by upper shift *O*, backspace, upper shift *M*), a monadic function: Rotate reverses the order along the indicated dimension. For example, ⌽*X* reverses the order of columns in *X*, an array of dimension 2 2 2.

⌽[2]*X* reverses the order of rows in the array *X*.

⌽[1]*X*, which reverses the order of planes, is equivalent to ⊖*X* (keyed by upper shift *O*, backspace, *MINUS*).

Rotate, ⌽, dyadic form: The amount and direction of rotation can be specified in the dyadic use of ⌽.

2⌽ (list) rotates a list two positions to the *left*.

¯3⌽ (list) rotates a list three positions to the *right*.

Dyadic rotate can be used in a manner analogous to the use of scan and reduction with higher-dimensional arrays.

⊖*X* is equivalent to ⌽[1]*X*.

Transpose, ⍉ (keyed by upper shift *O*, backspace, \ ), a monadic function: ⍉ applied to an array with dimensional elements [*I*; *J*; *K*] yields an array with dimensional elements [*K*;*J*;*I*].

Transpose, ⍉, dyadic form: If *A* is an integer array, *A*⍉*X* transposes the elements of *X* as specified by the elements of *A*. For example, if *X* has dimensions 2 3 4, then 2 3 1 ⍉ *X* produces an array with 3 planes, 4 rows, and 2 columns.

Ravel, ",'' a monadic function: converts any array into a list by listing the elements along the dimensions, starting with the last and working forward to the first.

Catenate (or Laminate), ",'' the dyadic form of ravel: The function is used by *Y*, [*A*]*X*, where *A* is the specified dimensional "index'' of a multidimensional array *X*. If *A* contains integers, the function is called catenate; if *A* contains nonintegers, the function is called laminate. With catenation, *Y* and *X* are joined together to form a larger array whose dimension along that indicated by *A* is the *sum* of the corresponding values for *Y* and *X*. The arrays must be conformable. With lamination a new dimension is created (see page 212).

Take, ↑, (keyed by upper shift *Y*): dyadic function which extracts elements from a list or array. If *B* has dimensions 2 3 4 ,1 2 3 ↑ *B* produces an array from the first plane of *B*, the first two rows, and first three columns of B. Similarly `¯1 2 ¯3 ↑ B` produces an array from the *last* plane of B, the first two rows, and *last* three columns of B.

Drop, `↓`, (keyed by upper shift *U*): dyadic function used in a manner analogous to take, `↑`, but drops (or deletes) elements from the array instead of taking them. What is obtained is what remains *after* the indicated elements have been deleted from the array.

Exercises

APL Practice

1. Let `X←2 4 4⍴ 32 ?32`. That is, X consists of two planes of 4 by 4 matrices each. Perform the following operations and examine the results with a view to understanding how to handle multidimensional arrays.

   (a) `+/X`
   (b) `+⌿X`
   (c) `+/[1]X`
   (d) `+/[2]X`
   (e) `+/[3]X`
   (f) `-/[1]X`
   (g) `-⌿X`
   (h) `+/[1]X[2;;]`
   (i) `+/+\X`
   (j) `+/+\[1]X`
   (k) `+/+\[2]X`
   (l) `2 2 2 ↑ X`
   (m) `0 0 1 ↑X`
   (n) `5 3 2 ↑X`
   (o) `1 1 1 ↓1 1 1 ↑X`
   (p) `+\+/X`
   (q) `0 0 0↓0 0 0 ↑X`
   (r) `+/ 1 1 0 ↓0 0 1 ↑X`
   (s) `1 0 0 1 /[2]X`
   (t) `1 0 1 0 /[3]X`
   (u) `(8⍴01)\[2]X`
   (v) `1,X`
   (w) `1,0,1,X`

2. Use the matrix X of exercise 1 to examine the following APL operations. Before carrying out each operation, try to predict the result.

   (a) `2⌽X`
   (b) `0⌽X`
   (c) `¯1⌽X`
   (d) `¯1⌽[1]X`
   (e) `¯1⌽[2]X`
   (f) `¯1⌽[3]X`
   (g) `⊖X`
   (h) `⍉X`
   (i) `1 2 3 ⍉X`
   (j) `2 1 3 ⍉X`
   (k) `3 1 2 ⍉X`
   (l) `⌽,X`
   (m) `1 0 ⌽,X`
   (n) `⌽X,[2]X`
   (o) `([2]X),⌽X`
   (p) `1,X`
   (q) `X,1`
   (r) `X;1`

3. Use the matrix X of exercise 1 to examine the difference between:

   (a) `¯1 ¯1 ¯1↓X` and `1 1 2 ↑X`

(b) `0 0 0 ↓X` and `2 3 4 ↑X`
(c) `+/[1]X` and `(1 3 4 ↑X)+1 0 0 ↓X`
(d) `X[;2;]` and `2 3 1 ↑1⌽X`
(e) `⊖X` and `⌽[1]X`

4. Use the matrix X of exercise 1 to :
   (a) Compute the left inverse of X.
   (b) Insert a row of zeros between the 3rd and 4th row of each block.
   (c) Insert two blocks of zeros between the two blocks of X.
   (d) Select the second block with a column of 1's added in the beginning.
   (e) Form a block-diagonal matrix consisting of the two blocks of X.
   (f) Form a matrix W of dimension 3×4 consisting of the 1st and 3rd rows of the first block of X and the 2nd row of the second block of X.

5. Define a matrix A with elements

$$a_{ij} = \begin{cases} (-1)^{j-1}\binom{j-1}{i-1} & i<j \\ (-1)^{i-1} & i=j \\ 0 & i>j \end{cases} \qquad j = 1 \ldots 10$$

where $\binom{j-1}{i-1}$ is the binomial coefficient. Verify that $A^2 = I$. The matrix A is said to be orthonormal; it is its own inverse!

6. For the matrix X of exercise 1, use the logical function, =, to show that

$$\begin{bmatrix} X[1;;] & 0 \\ 0 & X[2;;] \end{bmatrix}^{-1} = \begin{bmatrix} X[1;;]^{-1} & 0 \\ 0 & X[2;;]^{-1} \end{bmatrix}$$

A matrix such as

$$\begin{bmatrix} X[1;;] & 0 \\ 0 & X[2;;] \end{bmatrix}$$

is called block diagonal, where $X[1;;]$ and $X[2;;]$ are the blocks.

7. Consider the square matrix

$$A = \begin{bmatrix} 1 & 2 & 3 \\ 5 & 1 & 5 \\ 4 & 8 & 12 \end{bmatrix}$$

Use the computer to verify :
   (a) That A is singular (i.e., has no inverse).
   (b) That the matrix e^A is not singular.
   (c) That $C = A + B$, where

$$B = \begin{bmatrix} 0 & 0 & 0 \\ 0 & 0 & 0 \\ 0 & 0 & 0.1 \end{bmatrix}$$

is nonsingular.

(d) That $(I + B)A$ yields a matrix identical to A except that the 3rd row has been replaced by the third row plus 0.1 times the 3rd column, while $A(I + B)$, where I is the identity matrix, has a similar effect upon columns.

8. For $A = \begin{bmatrix} 1 & 0 \\ 0 & 2 \end{bmatrix}$ and $B = \begin{bmatrix} 3 & 0 \\ 0 & 1 \end{bmatrix}$ show that $AB = BA$ using the logical function '='.

9. Given the matrices:

$$Y = \begin{bmatrix} -1 & -1 \\ 1 & 0 \end{bmatrix} \qquad X = [X_{ij}]_{N \times N}$$

where $X_{ij} = (-1)^{N-j}\binom{N-j}{i-1}$, $\binom{N-j}{i-1}$ is the binomial coefficient, $N = 10$, $j = 1, \ldots, 10$, $i = 1, \ldots, 10$, find the limits of

(a) $\lim_{n \to \infty} Y^n$

(b) $\lim_{n \to \infty} X^n$

10. Assume that the array named ALL consists of 12 matrices, one for each of the 12 cities in which a publishing company sells some of its magazines. Assume further that each matrix consists of five rows that represent locations within the city, and that each row has 16 elements representing the number of magazines sold per month by each of the 16 newsstands in each location. How would you write in APL a program to determine:

(a) The total number of magazines sold in each city during the month.

(b) The newsstand with the maximum sales volume in each city.

(c) The total volume of sales.

(d) Let P be the row of the 12 different prices charged, one for each city. Show how you would find the city with the maximum sales in dollars.

Statistical Applications

Data for these applications are in appendix E.

1. Find estimates of the coefficients of the regression

$$I_t = a_0 + a_1 RL_t + u_t$$

where I_t = level of investment in the U.S. for the years 1950 to 1978, and RL_t = the long-term interest rate for the U.S. for the same 29-year period. Recall that if you use REGRESS, the data arrays must be lists, not two-dimensional arrays.

2. Suppose you wanted to rerun the regression of exercise 1, $I_t = a_0 + a_1 RL_t + u_t$, using the information (which somehow you managed to get) that the true variance of u_t is 225. How will this new information change your regression coefficient estimators and estimates? How about their estimated standard errors?

12

Inner and Outer Products—Matrix Manipulation

The previous chapter gave us a variety of methods for rearranging, selecting, and building up multidimensional data arrays. But multidimensional arrays also enable us to perform with ease a number of complicated statistical computations. For example, we frequently require the sum of products of observations on two random variables. Another frequent requirement is the preparation of tables and the need to perform some operation on all the elements of an array with each element of another array. The functions below facilitate these operations.

12.1 Inner Product: Some New Ideas

Inner Product
+.×

You have already seen the inner product in connection with matrix multiplication. In that case two matrices, say A and B, were operated upon using the three symbols that form what is sometimes called the "plus-times" inner product. We form two matrices

```
      A←2 2ρι8
      B←2 2ρ4+ι8
            A
  1  2
  3  4
            B
  5  6
  7  8
```

The plus-times inner product is written as

```
      A+.×B
19 22
43 50
```

Inner Product Generalized

To review, the (1, 1) entry is formed by $(1 \times 5) + (2 \times 7) = 19$; the (1, 2) entry by $(1 \times 6) + (2 \times 8) = 22$; (2, 1) by $(3 \times 5) + (4 \times 7) = 43$; and (2, 2) is computed by $(3 \times 6) + (4 \times 8) = 50$. We have, in effect, placed the multiplication symbol between the elements "across" the second coordinate of A and down the first coordinate of B. Then the summation sign is inserted between these product pairs.

Suppose that we reversed the + and × signs? We would have

```
      A×.+B
54  70
88 108
```

The (1, 1) element is computed by $(1 + 5) \times (2 + 7) = 54$, and the (1, 2) element is $(1 + 6) \times (2 + 8) = 70$. So you can see that the "inner product" is more powerful than you might have expected. Another example is maximum-times inner product:

```
      A⌈.×B
14 16
28 32
```

The first element is computed by `(1×5)⌈(2×7)=14`. Another function might be called the max-min inner product:

```
      A⌈.⌊B
2 2
4 4
```

where the first element (1, 1) is computed by `(1 ⌊ 5) ⌈ (2 ⌊ 7)=2`. For example, suppose that A and B represent expenditures on a group of commodities over time by two individuals. You wish to discover what was the largest of the minimum expenditures between the two.

These operations can be performed on higher-order arrays of different dimensions. For example:

```
      A←2 2 2 ρι8
      B←2 2 ρ1
      A
 1  2
 3  4
```

```
 5  6
 7  8
      B
 1  1
 1  1
      A+.×B
 3  3
 7  7
11 11
15 15
```

The shape of the result of this command is 2 2 2. The shape always drops the *last* coordinate of the array on the *left* and the *first* coordinate of the array on the *right*. The "dropped" coordinates must be equal to each other—this is called conformability of the two arrays. Just as before, the multiplication symbol is placed between the elements of the last coordinate of A and the first of B.

$$(1 \times 1) + (2 \times 1) = 3 \qquad (1 \times 1) + (2 \times 1) = 3$$
$$(3 \times 1) + (4 \times 1) = 7 \qquad (3 \times 1) + (4 \times 1) = 7$$

$$(5 \times 1) + (6 \times 1) = 11 \qquad (5 \times 1) + (6 \times 1) = 11$$
$$(7 \times 1) + (8 \times 1) = 15 \qquad (7 \times 1) + (8 \times 1) = 15$$

Another operation is illustrated by the following example. If we wanted to obtain the matrix product of the matrix in plane one of A with that of the matrix in plane two of A, we would key in

```
      A[1;;]+.×A[2;;]
19 22
43 50
```

This is exactly what we obtained in the first example of this section. If the two planes represent observations on two vectors of random variables, each observed over time, then the above operation produces the raw moment matrix of the random variables. Thus the typical (i, j) element of the inner product is mathematically $\Sigma_k x_{ik} y_{jk}$, where x_{ik} and y_{jk} are the kth observations on x_i and y_j, the ith and jth elements, respectively, in the x and y vectors of random variables.

12.2 Outer Product

Outer Product
∘.*fn*.g

This instruction enables you to place a primitive dyadic function between each element in the corresponding positions in the two arrays. (Is that a mouthful of words! And only a few weeks ago you didn't even know that APL means A Programming Language!) This function is constructed with two symbols, the null or jot (the little ° above the J) and the period. Its general form is

```
A∘.fn B
```

where *A* and *B* can be either scalars or matrices. ∘. is called "jot dot."

One use that is made of this instruction is preparing tables. Here is an example of a simple arithmetic table for addition.

```
      1 2 3 4 5∘.+0 1 2 3 4 5
1 2 3 4 5 6
2 3 4 5 6 7
3 4 5 6 7 8
4 5 6 7 8 9
5 6 7 8 9 10
```

The first row is computed by adding to one (the first element on the left) all the elements on the right of the outer product function. The second row is formed by adding the *second* element of the left argument to each element of the right argument. Here is another way to visualize the operation:

+	0	1	2	3	4	5
1	1	2	3	4	5	6
2	2	3	4	5	6	7
3	3	4	5	6	7	8
4	4	5	6	7	8	9
5	5	6	7	8	9	10

Notice that we have written in the column and row headings and the addition symbol. Let's try another simple arithmetic table, only this time let *B* be a three-dimensional array.

```
      B←2 2 2⍴¯4+⍳8
      B
¯3 ¯2
¯1  0

 1  2
 3  4
      C←0 1 2∘.×B
      C
```

```
 0   0 )
 0   0 |
       | Block 1, given by 0 × B
 0   0 |
 0   0 )

¯3  ¯2 )
¯1   0 |
       | Block 2, given by 1 × B
 1   2 |
 3   4 )

¯6  ¯4 )
¯2   0 |
       | Block 3, given by 2 × B
 2   4 |
 6   8 )
      ⍴C
3  2  2  2
```

Notice that the shape that results from the outer product is the *catenation* of the shapes of the arrays A and B, ⍴ (A) = 3, ⍴ (B) = 2 2 2. C has 3 blocks of 2 planes of 2 rows by 2 columns. The arithmetic is just as easy as it is in the simple case. The scalar zero is multiplied by every element of the three-dimensional array B, then $1 \times B$ is formed, and then $2 \times B$. Feel free to experiment with other dyadic operators or combinations of operators. However, be aware of how much storage, printing, and computer time you are using, since this is a "super-powerful" command. Consider how many values A∘.×B would generate if the shapes of A and B were each (5 5 5 5).

12.3 An Economic Example (Production Functions)

An interesting use of outer product concerns production functions. Suppose that we have estimated the parameters of a production function and we want to examine the properties of the estimated function. We start with a modified version of a Cobb-Douglas production function:

$$Y_i = \delta L_i^{\alpha} K_i^{\beta} e^{\gamma D_i} e^{U_i}$$

and our estimates of the parameters could have been

$$\hat{Y}_i = 120.0\ L_i^{0.5} K_i^{0.5} e^{0.3 D_i}$$

Y_i might be the amount of oil pumped from various drilling platforms, L_i equals the amount of labor at each platform, K_i is the amount of capital services at each of the platforms, and D_i is a binary variable that takes on the value 1 for offshore rigs and 0 for onshore rigs. $\hat{Y}$ is the estimate of the conditional mean of Y_i. The Greek letters are symbols for the parameters that were estimated, and U_i is a disturbance term distributed normally with zero mean and constant variance.

Our objective is to explain the economic implications of this equation. We write an APL program using the outer product command.

```
      ∇PROD
[1]  Y←120×(*0.3×D)∘.×(L*0.5)∘.×(K*0.5)
[2]   ∇
```

Working from right to left we raise K to the 0.5 power. Next perform outer product multiplication with L raised to the 0.5 power. This two-dimensional result is combined with $e^{0.3D}$ using the outer product multiply again. The final three-dimensional array is multiplied by 120.

Notice that we did not code the equation exactly as it was first written. Since the *first* dimension of the result is the coordinate of the *left*-most variable, we put D on the left. This allows us to see most easily the effect of drilling location on productivity. We generate some hypothetical data:

```
      L←⍳10
      K←⍳5
      D←0 1
      PROD
      Y
120          169.7056275  207.8460969  240          268.3281573
168.7056275  240          293.9387691  339.411255   379.4733192
207.8460969  293.9387691  360          415.6921938  464.7580015
240          339.411255   415.6921938  480          536.6563146
268.3281573  379.4733192  464.7580015  536.6563146  600
293.9387691  415.6921938  509.1168825  587.8775383  657.267069
317.4901573  448.9988864  549.9090834  634.9803147  709.929574
339.411255   480          587.8775383  678.8225099  758.9466384
360          509.1168825  623.5382907  720          804.9844719
379.4733192  536.6563146  657.267069   758.9466384  848.5281374

161.9830569  229.078636   280.5628845  323.9661138  362.2051265
229.078636   323.9661138  396.7758364  458.1572719  512.2354022
280.562845   396.7758364  485.9431707  561.1257691  627.3576818
323.9661138  458.1572719  561.1257691  647.9322276  724.4102529
362.2051265  512.2354022  627.3576818  724.4102529  809.9152845
396.7758364  561.1257691  687.2359079  793.5516728  887.217742
```

```
428.5668852  606.0851014  742.2996196  857.1337704  958.3046882
458.1572719  647.9322276  793.5516728  916.3145438 1024.470804
485.9491707  687.2359079  841.6886536  971.8983415 1086.615379
512.2354022  724.4102529  887.217742  1024.470804  1145.39318
```

The array *Y* has too many digits in each number for us easily to see what is happening. Let's get an array of approximations to *Y* that will be easier to read. Let's try

```
      YI←⌊Y
      YI
120 169 -  -  -
168 240 -  -  -
207 293 -  -  -
240  -
 -   -
 -   -
 -   -
```

This array has the following form:

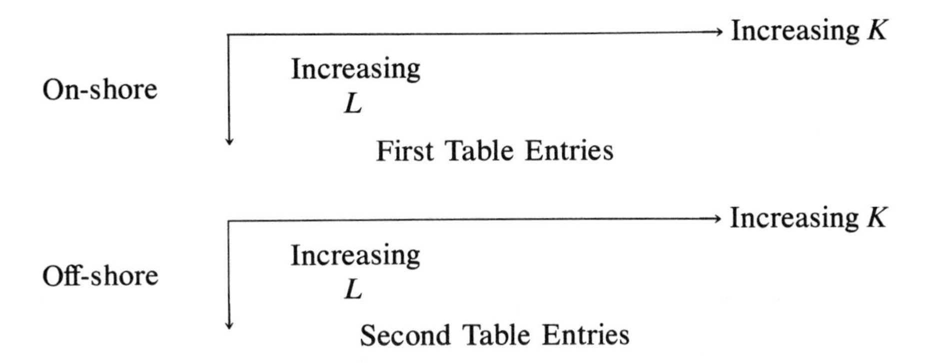

You might also want to see the effect that an increase in the output elasticity of capital from 0.5 to 0.7 would have on output with various combinations of inputs. Consider

```
      ∇PROD[1]
[1]  [1⎕]
[1]  Y←120×(*0.3×D)∘.×(L*0.5)∘.×(K*0.5)
[1]  Y←120×(*0.3×D)∘.×(L*0.5)∘.×(K*0.7)
[2]   ∇
```

```
      PROD
      ⌊Y
120  194   258   316   370
169  275   366   447   523
207  337   448   548   641
240  389   517   633   740
268  435   578   708   827
293  477   634   775   906
317  515   685   837   979
339  551   732   895  1047
360  584   776   950  1110
379  616   818  1001  1170


161  263   349   427   499
229  372   494   604   706
280  455   605   740   865
323  526   699   854   999
362  588   781   955  1117
396  644   856  1047  1224
428  696   924  1130  1322
458  744   988  1209  1413
485  789  1048  1282  1499
512  832  1105  1351  1580
```

These data can be plotted to give a clearer picture of the results.

You have seen to some extent how higher-order arrays can be used in APL to simplify your statistical calculations. However, it is easy to become confused at first. It will help you to avoid confusion if you experiment with small samples of data and check your work at each stage. We have kept the examples small so that they were relatively easy to check by hand. The last example on the production function was more ambitious, but now you are at a point where you can probably use APL to check your APL. Applications that are even more ambitious, but more useful, will be discussed next.

12.4 Two More Not-So-Elementary Matrix Operations (Kronecker Product, Determinant)

* Kronecker Product

Kronecker Product

Later on, if you continue to study econometrics and statistics, you will find a great need for a mathematical operation called the Kronecker product. The Kronecker product of two matrices Σ and W is written mathematically as $\Sigma \otimes W$, where Σ and W are square matrices of dimensions $(n \times n)$ and $(m \times m)$, respectively. The result, say C, is of dimension $(nm \times nm)$, and is defined by

$$C = \Sigma \otimes W = \begin{bmatrix} \sigma_{11}W & \sigma_{12}W & \cdots & \sigma_{1n}W \\ \sigma_{21}W & \sigma_{22}W & \cdots & \sigma_{2n}W \\ \cdot & \cdot & & \\ \cdot & \cdot & & \\ \cdot & \cdot & & \\ \sigma_{n1}W & \sigma_{n2}W & \cdots & \sigma_{nn}W \end{bmatrix}$$

where Σ and W are the matrices

$$\Sigma = \begin{bmatrix} \sigma_{11} & \sigma_{12} & \cdots & \sigma_{1n} \\ \sigma_{21} & \sigma_{22} & \cdots & \sigma_{2n} \\ \cdot & \cdot & & \\ \cdot & \cdot & & \\ \cdot & \cdot & & \\ \sigma_{n1} & \sigma_{n2} & \cdots & \sigma_{nn} \end{bmatrix} \qquad W = \begin{bmatrix} w_{11} & \cdots & w_{1m} \\ w_{21} & \cdots & w_{2m} \\ \cdot & & \\ \cdot & & \\ \cdot & & \\ w_{m1} & \cdots & w_{mm} \end{bmatrix}$$

and $\sigma_{ij}W$ represents *scalar* multiplication of the matrix W by the scalar σ_{ij}.

This may at first glance look like a complicated product to obtain, but by now you know that in APL the required computation will be easy. We use the outer product introduced earlier in this chapter.

Consider first the straightforward multiplicative outer product `A∘.×B`, which multiplies each element of the matrix A by each element of matrix B. So far so good. But the dimensions of the result are not exactly what are wanted. If D is the result of `A∘.×B`, where A is $(n \times n)$ and B is $(m \times m)$, then D has dimensions $(n\ n\ m\ m)$, that is, n blocks of n planes of m rows and m columns, or we have n-squared $(m \times m)$ matrices. You may recall that the shape of `A∘.×B` is the catenation of the shapes of A and of B. But we want an $(mn \times mn)$ matrix. The desired rearrangement of the result can be achieved by using reshape and the dyadic transpose, ⍉ (key upper shift `O`, backspace, key \).

As you will recall, the monadic transpose, when applied to a matrix, transposes rows and columns, but what about this strange beast of a multidimensional array D? `⍉D` would merely give us m blocks of m planes of $n \times n$ matrices; the order of dimensions is merely reversed. Thus, to get what we want, we need to do two things: rearrange blocks, planes, rows, and columns so that then we can reshape the result to get an $(mn \times mn)$ matrix.

Let's consider the problem of getting $A \times B$, where

$$\underset{(2\times 2)}{A} = \begin{pmatrix} 1 & 2 \\ 3 & 4 \end{pmatrix} \qquad \underset{(3\times 3)}{B} = \begin{bmatrix} 1 & 2 & 4 \\ 1 & 3 & 3 \\ 1 & 4 & 2 \end{bmatrix}$$

First, let's see what the outer product gives us (assign values to A and B first!):

```
1  2  4 }
1  3  3 }   this is equivalent to a11B  }
1  4  2 }                               }
                                        }   Block 1
2  4  8 }                               }
2  6  6 }   this is equivalent to a12B  }
2  8  4 }

3  6 12 }
3  9  9 }   this is equivalent to a21B  }
3 12  6 }                               }
                                        }   Block 2
4  8 16 }                               }
4 12  4 }   this is equivalent to a22B  }
4 16  8 }
```

If we try to reshape the above as it stands, we get the wrong results; try it and see. The solution we want is obtained by use of the dyadic transpose. Our solution is given by recalling that if we label the positions in A∘.×B by [I; J; K; L], we want to rearrange them so that we get instead [I; K; J; L], to get 2 blocks of 3 planes of 2 rows by 3 columns, which can now be reshaped into a (6 × 6) matrix. Consider then

```
1 3 2 4 ⍉D
```

If you type this into the computer and look at the output, you will observe that raveling the result (which proceeds by raveling the first matrix in the first plane in the first block, then the second matrix in the first plane in the first block), you will get an array that can be reshaped into that required for the Kronecker product. Let's see it.

```
      6 6 ρ1 3 2 4 ⍉D

1  2  4  :  2  4  8
         :
1  3  3  :  2  6  6
         :
1  4  2  :  2  8  4
         :
.........:.........
         :
3  6 12  :  4  8 16
         :
3  9  9  :  4 12 12
         :
3 12  6  :  4 16  8
```

The dotted lines have been inserted to aid you in relating the output to the definition of a Kronecker product.

Dyadic transpose gives us another advantage—an easy way to get the diagonal elements of a matrix. Let K be defined by

```
      K←6 6 ρ1 3 2 4 ⍉D
```

Try

```
      1 1 ⍉K
1 3 2 4 12 8
```

In short, using `1 1 ⍉` is how to get the diagonal elements of a matrix.

The trace of a matrix is the sum of its diagonal elements, so that in APL the trace is

```
      +/1 1 ⍉K
30
```

Determinant

Determinant

The next matrix operation you may need is the determinant, written as $|A|$. The determinant of a square matrix A of dimension $(n \times n)$ can be written as a linear function of determinants of submatrices of A of dimension $(n - 1) \times (n - 1)$; in short, the determinant can be defined for $n = 1$ by A itself. For $n = 2$, $|A|$ is given by $(a_{11}a_{22} - a_{12}a_{21})$. For larger n, if we let $\hat{A}_{ij} = (-1)^{i+j}D_{ij}$, where D_{ij} is the determinant of the matrix obtained from A by deleting the ith row and jth column, and let a_{ij} be the (i, j)th element of A, then

$$|A| = \Sigma_{j=1}^{n} a_{ij}\hat{A}_{ij}$$

for any $i = 1, 2, \ldots, n$. This is the usual expansion by minors, which we can program quite easily in APL. Let's begin by writing things out using paper and pencil.

We see that the routine for calculating the determinant can be decomposed into two main components: the calculation of the determinant of a 2×2 matrix, and the recursive definition of a determinant of a matrix of dimension n in terms of determinants of matrices of dimension $(n - 1)$. So the basic elements of the routine are these two components, plus a decision component that enables the computer to know which of the other two components to calculate.

The determinant of a 2×2 matrix is easily programmed. Consider

```
      D←(×/1 1 ⍉A)-A[1;2]×A[2;1].
      D
¯2
```

which is nothing more than the APL version of $(a_{11}a_{22} - a_{12}a_{21})$. You will recall that `1 1 ⍉ A` gives the diagonal elements of A, so adding `×/` yields the product of the diagonal elements.

If E is a vector containing the signs associated with the minors D_{1j}, that is, E is a vector with elements $1 \; -1 \; 1 \; -1 \ldots$ generated by $(-1)^{i+j}$ for $j = 1, 2, \ldots, n$, then the recursive definition is

```
D←A[1;] +.× (E×M)
```

which is the APL version of

$$\Sigma_{j=2}^{n} a_{1j}(-1)^{1+j} D_{1j} = \Sigma_{j=1}^{n} a_{1j} \hat{A}_{1j}$$

Looping

Looping

We note from our definitions that the vector M has to be created element by element by getting the determinant of various submatrices. In short, to evaluate the function determinant, our function has to call itself! Fortunately, in APL this causes no difficulties, provided that it is done correctly. We also see that we have to do the same thing $(n - 1)$ times (for an n-dimensional matrix) each time we call the function; this is called "looping."

In order to define M, we have to calculate the determinants of the matrices $A_{11}, A_{12}, \ldots, A_{1n}$, where A is the matrix which is the argument of our function. A_{1j}, you will recall, is obtained from A by deleting the *first row* and the jth *column*. To do this in APL, we will need to use the indexing of arrays and the "drop," `↓`, operation (key upper shift `U`). Another operation we need in this routine is "compression," which is the dyadic use of /.

In our problem we want to drop only one row (the first) and one column, but a different one each time; this is where the indexing comes in. If we rearrange the matrix A so that the column to be dropped is first, and if we keep everything else in the same order, you can easily see (at least by using a paper and pencil to try it) that this will give us the submatrices we want. Consider the following:

```
      N←1 0/ ρA                 {We need to know how "big" A is.

      IND←⍳N                    {Gives us the index numbers 1,
                                {2, . . . , N.
      )ERASE B

      B←(N,N) ρ0

      B
                                {We need a "dummy" matrix for
0 0                             {reindexing.

0 0

      I←1                       {Initializes our index.
                                {Produces an index array
      IN←(I≠IND)/IND            {with the Ith element deleted.
      M←M, DET 1 1 ↓B[;IND]←A[;I,IN]
```

The last line needs some discussion. First, let's call our determinant function *DET* and let it have one argument, the matrix whose determinant we are trying to obtain. Also, by writing `M←M,DET` (argument), we create a list *M* with determinants as its elements by *catenation.*

```
B[;IND]←A[;I,IN]
```

rearranges the columns of *A* so that the column to be deleted, indexed by *I*, occurs first in *B*, and the remaining columns are unchanged in their relative positions. `1 1↓B` deletes the first row and column of *B*, and *DET* takes the determinant; in short, we get the (1, *I*) minor by this line.

We are nearly done—the next most important step is to complete the looping through the index *I* from 1 to *N*. This is done as follows:

```
I←1                                  {Initializes the loop.
ST:→ST1×ι(I > N)                     {Gets us out of the loop
                                      when I exceeds N.
IN←(I≠IND)/IND
M←M,DET 1 1 ↓B[;IND]←A[;I,IN]
E←E, (¯1*(1+I))                      {Produces the vector of signs.
I←I+1                                {Increments I for the looping.
→ST                                  {Takes us back to the be-
                                      ginning to check on I.
```

We now have only a few minor details to settle: the original decision of whether to go to the $n = 2$ case and some "housekeeping chores." For example, in order to define *M* and *E* in the way in which we have done it, they have to be "defined as lists," even if there are no elements in them. This is done by setting $M \leftarrow \iota 0$ and $E \leftarrow \iota 0$.

The whole routine is

```
      ∇D←DET A;N;E;I;IN;IND;M;B
[1]   N←1 0 / ρA
[2]   →END×ι(N=2)
[3]   IND←ιN
[4]   B←(N,N)ρ0
[5]   I←1
[6]   M←ι0
[7]   E←ι0
[8]   ST:→ST1×ι(I > N)
[9]   IN←(I≠IND)/IND
[10]  M←M,DET 1 1 ↓B[;IND]←A[;I,IN]
[11]  E←E,(¯1)*(I+1)
```

```
[12] I←I+1
[13] →ST
[14] ST1:D←A[1;]+.×(E×M)
[15] →0
[16] END:D←(×/ 1 1 ⍉A)-A[1;2]×A[2;1]
     ∇
```

Let's try it.

```
      A←2 2 ρ1 2 3 4
      DET A
¯2
      Z←3 3 ρ1 2 3 2 3 4 3 4 5
      DET Z
0
      D←5 5 ρ3 7 1 2 5 6 4 3 0 2 0 3 0
              1 2 1 0 6 5 3 2 1 0 2 0
      DET D
278
```

A Routine to Calculate the Determinant of a Matrix

With this last one, if you waited around for what seemed like an age (actually, about 3 minutes if you are on the IBM 5110—though only a few seconds on a large computer), you are right in concluding that the calculations seem to be *very* slow. This example raises a very important issue: the straightforward programming of a mathematical statement need not be the most *computationally* efficient way to calculate something. The main problem with the above routine is that the calculation of all the subdeterminants involves a very large number of operations—a number that increases with the dimension of A very rapidly indeed.

A computationally more efficient way of obtaining the determinant is to convert A to a triangular matrix (e.g., all entries below the diagonal are zero), and then the determinant is simply the product of the diagonals. Here is a more efficient routine.*

```
      ∇ D←PDET A
[1]   N←(ρA)[1]        ←An alternative to getting dimension of A.
[2]   I←1              ←Initializes I.
[3]   B←A[;I]          ←Stores Ith column of A in B.
[4]   B[⍳I]←0          ←Zeros out the first I elements of B.
```

* This routine was written by our colleague Dr. Richard W. Parks, University of Washington, Seattle.

```
[5]  A←A-(B∘.×A[I;])÷A[I;I]        ←See below.
[6]  →(N≥I←I+1)/3                  ←See below.
[7]  D←×/ 1 1 ⍉A                   ←Obtains determinant by multiply-
                                     ing the diagonal elements.
     ∇
```

Only lines [5] and [6] require any explanation. The main idea of this routine's algorithm (method of calculating a mathematical formula) occurs in line [5]. The idea is to transform the matrix A into a triangular matrix; in this case all elements below the diagonal are zero. This is accomplished by subtracting from each row of A the elements of A to the left of the diagonal.

Line [6] increments the index I and then returns to line [3] if $N \geq I$; otherwise we continue to line [7] to calculate D.

```
      PDET Z
0
      PDET D
278
```

If you try this routine, you will find that it will execute the matrix D much faster than does the previous routine. Indeed, on the IBM 5120 it will do the D matrix in about 7 seconds. However, even this routine has its problems, since it will not work if any of the diagonal elements are zero.

While this lesson may have been a bit expensive, the lesson to be learned is important. As your routines become more complex, you will have to worry about computationally efficient ways of handling the problem. Finally, you should realize that the usual mathematical statements, while very informative, can often be computationally inefficient.

Summary

Inner Product and Multidimensional Arrays: the general form of this product is $A fn_1 \cdot fn_2 B$, where A, B are arrays, and fn_1, fn_2 are functions. For example, `A+.×B` where `A`, `B` are two arrays, gives the usual mathematical operation of "matrix product." The inner product for multiple dimensional arrays is always between the elements of the *last* dimension of A and the elements of the *first* dimension of `B`.

Outer Product, ∘.*fn* (keyed by upper shift *J*, period, followed by a function): general use is given by A ∘.*fn* B where A, B are arrays. The output is an array whose dimensions are the *catenation* of the dimensions of A and of B. Each element of A is an argument to the dyadic function *fn* for every element of B.

Kronecker Matrix Product: mathematically is given by $\Sigma \otimes W$, where Σ and W are $(n \times n)$ and $(m \times m)$ matrices, respectively. The result is of dimension ($nm \times nm$) and each element of Σ scalar multiplies the matrix W. The Kronecker product, K, for two matrices A, B is obtained in APL by

```
MN←(N←1 0/ ⍴A)×(M←1 0/ ⍴B)

K←MN MN ⍴1 3 2 4 ⍉A∘.×B
```

Determinant of a square matrix A, denoted by $|A|$: defined mathematically by

$|A| = \Sigma_{j=1}^{n} a_{ij}\hat{A}_{ij}$,   for any $i = 1, 2, \ldots, n$.
a_{ij} = (i, j)th element of A
$\hat{A}_{ij} = (-1)^{i+j} D_{ij}$
$D_{ij} = |A_{ij}|$, called the (ij) minor of A.
A_{ij} = matrix A with ith row and jth column deleted, is the (ij)th cofactor of A.

Two illustrative APL routines are given in the text; care is needed in the use of both.

Exercises

APL Practice

1. For any list L, the APL expressions `L⌹((⍴L), 1)⍴1` and `L⌹L÷L` give the same number. What is this number? Why is this result true?
2. For two positive numbers A and B, the APL expressions `((A*2)+B*2)*.5` and `A×4○B÷A` give the same number. What is this number? What is the explanation?
3. If `X←○(0, ⍳12)÷12` and `Y←⍉ 1 2 3 ∘.○ X`, what is in Y?
4. Let M be any matrix and `L←(,M>K)/,M` where K is any number. What is in L?
5. Let `Q←?N ⍴ K` and `F←+/[1]Q∘.=⍳K` for any two positive numbers N and K. What is in `F`?
6. Use the outer product to construct the following tables.
   (a) A table 100 by 10 of the powers from 1 to 10 of the integers from 1 to 100.
   (b) A table 100 by 2 of the logarithms to base e and base 10 of the integers from 1 to 100.
   (c) A table 100 by 10 of the 1st through the 10th root of the integers from 1 to 100.
   (d) A table of all possible products of the integers from 1 to 30.
   (e) A table of the binomial coefficients $\binom{j-1}{i-1}$ for $i < j$ and $j = 30$. (See exercise 9, Chapter 11.)

(f) The following matrix:

$$\Sigma = \begin{bmatrix} 1 & \rho & \rho^2 & & \cdots & \rho^{T-1} \\ \rho & 1 & \rho & \rho^2 & \cdots & \rho^{T-2} \\ \cdot & \cdot & \cdot & \cdot & \cdot & \cdot \\ \cdot & \cdot & \cdot & \cdot & \cdot & \rho^2 \\ \cdot & \cdot & \cdot & \cdot & 1 & \rho \\ \rho^{T-1} & \cdot & \cdot & \cdot & \rho & 1 \end{bmatrix}$$

for $T = 30$, where ρ can take the values 0.2, 0.5, 0.7.

7. Here is how the outer product can be used to plot functions. Consider the following steps:

(a) `Y←⌽X←¯11+⍳21`

(b) `L←Y∘.×X`

(c) Let `W←Y∘.=X+1`

(d) `K←' +'[1+0=W]`
  (one blank space)

(e) `PLOT←' ∇+∇'[W+1+2×0=L]`
  (one blank space)

Use this procedure to plot the following functions:

(a) $Y=2X+1$

(b) $Y=X^2-1$

(c) $Y=-3X^2+2X+3$

8. Since you know that the rank of an idempotent matrix is equal to its trace, find the rank of $M = I - X(X'X)^{-1}X'$, where

$$X = \begin{bmatrix} 1 & 4 & 1 \\ 2 & 1 & 6 \\ 3 & 5 & 2 \end{bmatrix}$$

9. The following matrix is called a payoff matrix of firm A which has only one competitor, firm B. Firm A has 3 possible pricing policies and firm B has 4 possible pricing policies. The elements of the matrix represent profit in \$1000 of firm A given B's possible pricing policies.

		B's Strategy			
		1	2	3	4
A's Strategy	1	50	90	18	25
	2	27	5	9	95
	3	64	30	12	20

There is another matrix for B which is not given because its elements are known to be the differences between the entries in A's matrix and

100 (assuming that only $100,000 profit can be made in the market). If A chooses strategy 1, then B will choose strategy 3, because B gets $82,000, which is the highest profit for B given A's strategy. Thus A's one optimum policy (called maximin) is to find the minimum of each row and then pick the maximum of those minima. This long verbal story is only half a line in APL. Can you write it? What if A chooses a *minimax* strategy? (A minimax strategy is exactly the opposite to the maximin strategy.)

10. Consider the following table:

Table M

		Possible home mortgage rate values this year in %			
		10.50	11.00	11.50	11.75
Possible home mortgage rate values next year in %	10.50	0.1	0.25	0.4	0.25
	11.00	0.25	0.3	0.25	0.20
	11.50	0.4	0.25	0.1	0.25
	11.75	0.25	0.20	0.25	0.3

Each element of this table represents the probability of next year's interest rate given the rate this year. Notice that column and row elements sum to 1. Enter the matrix of probabilities into the variable M.

(a) Find next year's expected interest rate if this year's interest rate is 11.50%.

(b) Find next year's expected interest rates under all possible alternative values for this year.

(c) Let

$$r_t = \begin{bmatrix} 10.5 \\ 11.0 \\ 11.5 \\ 11.75 \end{bmatrix}$$

Then Mr_t will be a vector of next year's expected values given this year's alternatives. Let's call this vector r_{t+1}. Then $r_{t+2} = Mr_{t+1}$ will be the potential values in year r_{t+2}.Give alternative interest rates for year $t + 1$. Project possible interest rates up to 5 periods ahead. Do they seem to converge to an equilibrium value?

11. Enter both routines *DET* and *PDET* in your workspace (see pages 233–235) and find the determinants of the matrices

(a) $A = \begin{bmatrix} 1 & 2 & 4 \\ 2 & 4 & 1 \\ 3 & 1 & 3 \end{bmatrix}$ (b) $B = \begin{bmatrix} 1 & 3 & 2 \\ 1 & 4 & 2 \\ 1 & 1 & 2 \end{bmatrix}$

(c) AB

(d) $A \otimes B$, where $\otimes$ is the Kronecker product.

(e) $B \otimes A$

12. Using the *PDETA* function (page 234), solve the following system of equations utilizing Crammer's rule, which is explained below:

$$2X + 3Y - 3Z = 7$$

$$X - 2Y + Z = {}^{-}2$$

$$3X + Y + Z = 9$$

Solutions by Crammer's rule are given by:

$$X = \frac{\Delta_X}{\Delta}, \qquad Y = \frac{\Delta_Y}{\Delta}, \qquad Z = \frac{\Delta_Z}{\Delta}$$

where

Δ is the determinant of the matrix of coefficients.

Δ_X is the determinant of the matrix Δ with the first column replaced by the column of constants.

Δ_Y is the determinant of matrix Δ with the second column replaced by the column of constants.

Δ_Z is the determinant of the matrix Δ with the third column replaced by the column of constants.

13. Let `C←¯1+?10 ⍴2`. Examine carefully the results of the following operations.

(a) `2∧C`

(b) `1∨C`

(c) `0∨C`

(d) `C+.∧C`

(e) `C∘.∧C`

(f) `C∘.⍱C`

(g) `C∘.⍲C` and compare to (e).

(h) `C∘.∨C` and compare to (f).

14. (a) Assign the statement "I don't like APL and tea" to variable X.

(b) Write an APL function that will erase the *n*, the apostrophe and the *t* of the word *don't* using the logical operators.

13

Linear Regression

For those of you who already know a fair amount of statistics, here is the chapter you have been waiting for. Now is the time to come to grips with regression analysis, the calculation of confidence intervals, and tests of hypotheses. In this chapter you will learn the advantages of being able to do all your statistical analysis yourself instead of relying on someone else's black box. Before you proceed, you might want to refresh your memory about simple linear regression (Chapter 9) and the use of ⌹ (Chapter 10).

This chapter deals only with the analysis of single-equation (as opposed to multi-equation), and linear (as opposed to nonlinear), regression equations. You might well be wondering why we have a whole chapter on regression, when the regression of a vector Y on a matrix X is obtained in APL by Y ⌹ X. The answer is that there are a number of associated statistics with a linear regression which need calculation and these are a little messier computationally. Let's begin.

13.1 Covariance and Correlation Matrices

Moment, Covariance and Correlation Matrices

The basic input to any regression problem is an $(N \times K)$ matrix of N observations on K regressors (variables used to ''explain'' the dependent or regressand variable Y). An important set of statistics for many reasons is the sample moment, covariance, and correlation matrices. If X is the

$(N \times K)$ regressor matrix, the mathematical statements are, in matrix terms,

Moment matrix	$X'X$
Covariance matrix	$X'X - \bar{X}'\bar{X}$
Correlation matrix	$D^{-1/2}(X'X - \bar{X}'\bar{X})D^{-1/2}$

where you may recall that $X'X$ is the inner product between X transpose and X itself, $\bar{X}$ is an $N \times K$ matrix, each of whose rows contains the array of means of the K regressors, and D is a diagonal matrix whose ith diagonal element is the variance of the ith regressor variable, so that $D^{-1/2}$ is the diagonal matrix of the inverse of the square roots of the nonzero elements of D.

The above mathematical equations give the matrix equivalents of the variances and covariances we calculated in Chapters 5 and 9. Each of these matrices is easily obtained in APL. They are, in turn,

```
Moment matrix:      MM←(⍉X)+.×X
Covariance matrix:  CM←MM-(XB∘.×XB←+⌿X)÷(⍴X)[1]
Correlation matrix: CRM←D+.×CM+.×D←⌹(((⍳K)∘.=⍳K←(⍴X)[2])
                                         ×CM)*0.5
```

Let's explain this. The moment matrix is easy: `⍉X` produces X transpose and `(⍉X) +.× X` gives the inner product of `⍉X` with X. The covariance matrix introduces something we used briefly in Chapter 9, namely `(⍴X)[1]`, which means: shape of X over the first dimension, in short, the number of rows. The rest is easy to understand once you realize that $\bar{X}'\bar{X}$, where $\bar{X}$ is an $(N \times K)$ matrix of the means of the K variables, is simply N (= number of rows = `(⍴X)[1]`) times the outer product of the array of means, or $1/N$ times the outer product of the vector of column sums of X.

Simple Correlation Matrix of Regressors

The correlation matrix is obtained from the covariance matrix by dividing the covariance between the ith and jth variables by the product of the standard deviations (square roots of the variances) of the ith and jth variables. The variance of the ith variable is clearly given by the ith diagonal term of CM. The elements of CRM, usually labelled in *mathematical* notation as $\rho_{ij} = \text{Cov}(x_i, x_j)/\sqrt{\text{Var}(x_i)\text{Var}(x_j)}$, satisfy the mathematical constraint $-1 \leq \rho_{ij} \leq 1$. For further mathematical and statistical details see Mendenhall and Reinmuth or Kmenta, both of which are listed in the bibliography at the end of the book.

The APL expression is now fairly obvious. K is the number of regressors. `((⍳K)∘.=⍳K)×CM` produces a diagonal matrix with the diagonal elements of CM, and `⌹(((⍳K)∘.=⍳K)×CM)*.5` gives the inverse of the square roots of the variance obtained from the CM matrix; the reason that we cannot use `*¯0.5` is that the zero off-diagonal terms will give `DOMAIN ERROR`. The operation `D+.×CM+.×D` produces a matrix with ρ_{ij} on the off-diagonal positions and 1's on the diagonal.

For example, let us suppose that we have the following X matrix in the computer's memory:

$$\underset{(12 \times 2)}{X} = \begin{matrix} 100 & 5.50 \\ 90 & 6.30 \\ 80 & 7.20 \\ 70 & 7.00 \\ 70 & 6.30 \\ 70 & 7.35 \\ 70 & 5.60 \\ 65 & 7.15 \\ 60 & 7.50 \\ 60 & 6.90 \\ 55 & 7.15 \\ 50 & 6.50 \end{matrix}$$

and if we don't, let's enter X now.

Consider a simple routine to calculate the moment matrices that we have been discussing.

```
      ∇ MAT[⎕] ∇

     ∇CRM←MAT X;D;MM;CM;K;XB

[1] 'MM EQUALS';MM←(⍉X)+.×X

[2] 'CM EQUALS';CM←MM-(XB∘.×XB←+⌿X)÷(ρX)[1]

[3] K←(ρX)[2]

[4] 'CRM EQUALS'

[5] CRM←D+.×CM+.×D←⌹(((ιK)∘.=ιK)×CM)*0.5

     MAT X

MM EQUALS

     61050          5577.5

        5577.5         544.2075

CM EQUALS

     2250          ¯54

      ¯54           4.857291667

CRM EQUALS

     1             ¯0.5165417262

      ¯0.5165417262  1
```

And if we have the following three-column regressor matrix:

```
     ZZZ
```

```
 55      100     5.5
 70       90     6.3
 90       80     7.2
100       70     7
 90       70     6.3
105       70     7.35
 80       70     5.6
110       65     7.15
125       60     7.5
115       60     6.9
130       55     7.15
130       50     6.5
```

Our routine produces:

```
      MAT ZZ
MM EQUALS
      126300          80450          8170.25
       80450          61050          5577.5
        8170.25        5577.5         544.2075
CM EQUALS
      6300           ¯3550            125.25
     ¯3550            2250            ¯54
        125.25        ¯54               4.85729166667
CRM EQUALS
      1               ¯0.942902569586  0.715995624984
     ¯0.942902569586   1               ¯0.516541726216
      0.715995624984  ¯0.516541726216   1
```

13.2 Some Initial Linear Regression Statistics

The first generalization of the regression model which was discussed in Chapter 9 is to allow for more than one regressor; to some extent we have done this already with the use of the function ⌹. The general model to be discussed in this chapter is

$$Y = XB + U$$

where Y is an $(N \times 1)$ regressand vector, X is the $N \times K$ matrix of regressors, B is the $K \times 1$ vector of regression coefficients to be estimated, and U is the $N \times 1$ vector of unobserved disturbance terms. Excellent elementary discussions of multiple regression analysis can be found in both of the references mentioned on page 241.

From our work in Chapters 9 and 10, you know the least-squares approach to estimating the regression coefficients involves the calculation of the left inverse of X. The left inverse of X is $(X'X)^{-1}X'$ (recall that A_{LI} is the left inverse of A if $A_{LI}A = I$) and in APL this is obtained by ⌹*X*, using the monadic use of domino. The least-squares regression coefficients are obtained by the dyadic use of domino, *Y*⌹*X*, which is the APL equivalent of the mathematical expression $(X'X)^{-1}X'Y$. Mathematically, we obtain from $(X'X)^{-1}X'Y$ the expression $B + (X'X)^{-1}X'U$. The statistical properties of the least-squares estimator depend on the statistical properties of the vector $(X'X)^{-1}X'U$. Let's store the regression coefficient estimators in *BE,* i.e., we compute *BE*←*Y*⌹*X*; *BE* stands for *B* vector *E*stimate.

As we mentioned in the note to instructors, some computer systems have not implemented the dyadic form of domino. An easy way around this is to write your own function directly.

```
      ∇Y DQ X
[1] (⌹((⍉X)+.×X))+.×((⍉X)+.×Y)∇
```

Here is how it works for a 3 independent variable case where the intercept is constrained to be equal to zero.

```
CLEAR WS
      Y←10?10
      X←10 3⍴30 ?30
      Y⌹X
0.076969 0.0021871 0.24941
      (⌹ ((⍉X)+.×X) )+.×((⍉X)+.×Y)
0.076969 0.0021871 0.24941
```

The estimator of the variance of the disturbance term was given for the special case examined in Chapter 9. The general mathematical expression is

$$\widehat{\text{Var}}(\hat{U}) = \Sigma_i(Y_i - \hat{Y}_i)^2/(N - K)$$

where $\hat{Y} = X\hat{B}$, $Y' = (\hat{Y}_1, \hat{Y}_2, \ldots, \hat{Y}_n)$, where ∧ symbolizes the *estimator* of the variable or parameter under the ∧. Thus $\hat{B}$ symbolizes the estimator of B in the model $Y = XB + U$, where $\hat{B} = (X'X)^{-1}X'Y$.

The maximum-likelihood estimator of the variance of U is a little different, involving division by N instead of $N - K$. $\hat{Y}_i$ is the forecast value of Y_i for the ith observation given the regression coefficient estimates, i.e., $\hat{Y}_i =$

$\Sigma_i^N X_{ij}\hat{B}_j$. $\hat{B}_j$, $j = 1, 2, \ldots, K$ are obtained in APL by calculating the vector *BE* defined above. The APL expression for the estimator of the variance of *U* is

```
NP← ρX
V ←(SSE←(Y-X+.×BE)*2)÷(-/NP)
```

and in the process, we have defined the "error sum of squares," *SSE*, which will be needed later. Note that *NP* is the array (N, K), and `-/NP` is simply ($N - K$). $\hat{Y}$ is obviously given by `X +.× BE`.

The estimated covariance matrix of the regression coefficient estimators has two forms, depending on whether or not the regression equation is run in "deviation terms" [that is, after elimination of the "constant vector" $i' = (1, 1, \ldots, 1)$ by subtracting the mean from each variable]. If the regression is run in deviation terms, the estimated covariance matrix for *BE* is given by

```
COVD ← V × CM
```

If the regression is not run in deviation terms, the estimated covariance matrix for *BE* is given by

```
I ← 1
RM ← (⍉Z) +.× Z ← I,X
COV ←V× RM
```

The dimensions of *COVD* are ($K \times K$) and of *COV*, $((K + 1) \times (K + 1))$, when there are K regressors in X. The use of the catenate function in creating the matrix Z from I and X was discussed in Chapter 11.

The coefficient of determination (R^2) was defined in Chapter 9; the multiple correlation coefficient is merely the square root of R^2. Note that the coefficient of determination defined in Chapter 9 by 1-*SSE*/*SST* gives the coefficient of determination with respect to *variables defined in terms of deviations about their respective means.* If the R^2 value is wanted in terms of the original variables, then the new R^2 is defined as before, except that *SST* is given by `Y+.×Y` instead of `Y+.×Y-((+/Y)*2)÷N` .

13.3 Simple and Partial Correlation Coefficients

Simple and Partial Correlation Coefficients

A topic related to regression analysis is the analysis of simple and partial correlation coefficients. The simple correlation coefficient has already been defined and calculated in this chapter. Given K variables, the simple correlation between any two of them, say the ith and jth, is given mathematically by $r_{ij} = \text{Cov}(x_i, x_j)/\sqrt{\text{Var}(x_i)\text{Var}(x_j)}$; the simple correlation coefficient "measures" the extent of linear association between the two variables i and j.

A related concept is the partial correlation coefficient, say $r_{ij \cdot k}$, which is the correlation between variables i and j *after allowing for the joint correla-*

tions between variables k and i and between variables k and j. The partial correlation coefficient can be obtained easily from the simple correlation coefficients; mathematically, one has

$$r_{ij \cdot k} = (r_{ij} - r_{ik}r_{jk})/\sqrt{(1 - r_{ik}^2)(1 - r_{jk}^2)}$$

where r_{ij}, r_{ik}, r_{jk} are the simple correlation coefficients.

The matrix *CRM* obtained above gives for any set of *K* variables the matrix of *simple* correlation coefficients between each of the *K* variables. Any partial correlation coefficient can be obtained from the *CRM* matrix. By now you should be able to program this on your own.

13.4 Creation of a Regression Routine

Let us now consider creating a reasonably complete regression routine. The procedure will not only provide us with our own regression routine which, of course, we can alter in any way at any time we like, but it also will provide some valuable lessons in how to write routines that are more complicated than those we have tackled so far. Let's begin at the very beginning, which is to determine what we are required to do mathematically. Our next step will be to plan our programming steps. Here then is a recommended procedure for you to follow in writing any program routine of any difficulty.

1. Write down all the mathematical expressions needed; sequence the expressions so that all operations to be completed before a particular step are in fact listed before that step. One diagrammatic way to help you do this is to write down a flow chart, an example of which is given for the regression routine below.
2. Figure out whether any loops or conditional branch statements are needed and, if so, where in the program they are needed. Rethink the basic APL approach to see if there are alternative ways to get results which will not use up inordinate amounts of computer time and space.
3. Decide for each function whether the result is to be explicit or implicit, have one or two arguments, and which variables, if any, are to be globally defined.
4. *Begin with paper and pencil* and assemble the various parts of the routine that may already be written out.
5. Using your flow chart, lay out the sequence of "one liners" and set up your decision branches. As you do so, *label* all statements involving branches.
6. With each "one liner" expression, start with the "core" of the mathematical term and put it in the middle of the page; build outwards from there, and then recheck by reading the expression from *right to left*.

7. Alter the header line as you introduce local variables to avoid forgetting them.

8. Recheck your program carefully, looking to see that:
   (a) program statements give the results desired (read from right to left!);
   (b) the relationships between arrays, vectors, and matrices are comformable (possible length errors);
   (c) every variable will be defined *before* it is used;
   (d) loops, branches, and recursive function definitions will terminate under all alternatives.

9. *Document* your routine very carefully and *include explanatory comments* inside it (within a month, you won't remember how you did it!).

10. Enter the routine into the computer, then enter several easy problems for which you have the answers in order to check that the routine is correct. Be sure that you check the routine for general, not just special, cases.

11. Remember, even when you believe it is *impossible* to make a mistake, you will!

You might well ask: Why all these elaborate instructions? Up until now most of our routines have simply been written out. The answer is that so far the routines have all been very uncomplicated—the regression routine to follow is our first example of a more elaborate routine. Also, it is the first routine that we might want to keep for future use, so we need to be more careful about how we define it and set it up for use by others.

The following routine is a suggestion; a lot of variables are global since it is assumed that the output will be used in further analysis. We need some trial data. Let us suppose that the first column of *ZZZ* is the regressand array *Y*, and the remaining two columns of *ZZZ* are the regressor matrix *X*. Thus

```
      Y←ZZZ[;1]

      X←ZZZ[;2 3]

      Y

55 70 90 100 90 105 80 110 125 115 130 130

      X

100    5.5

 90    6.3

 80    7.2

 70    7

 70    6.3
```

```
70    7.35
70    5.6
65    7.15
60    7.5
60    6.9
55    7.15
50    6.5
```

Let us begin to write out the regression routine by deciding on what it is we want.

Desired Calculations

regression coefficient estimates
raw moment matrix
covariance matrix for regression coefficients
variance of disturbance term U,
R^2 (the coefficient of determination)
correlation matrix for regressors
array $\hat{Y}$
array $\hat{U} = (Y - \hat{Y})$
F statistics with degrees of freedom
standard deviations of the regression coefficient estimates
T ratios for the regression coefficient estimates
Disposition: all of the above to be global variables; but $\hat{Y}$, $\hat{U}$ to be printed out automatically; two arguments Y and X.

An examination of the formulas in Table 13.1 shows that there is a certain order in which calculations should be made and certain intermediate products which need to be defined as they are reused repeatedly. While the mathematical structure of this routine is very simple and hardly calls for an elaborate flow chart, one is given in Figure 13.1 in order to illustrate the idea. Our main concern is to keep the number of computations and numbers of arrays to a minimum, the former to economize on computer time, the latter to economize on space taken up in the computer.

In the process of working up the flow chart, you can see that some ideas occurred to us. First, we should remember that a function or a routine is defined for general use, so you should be on your guard to watch out for problems caused by someone trying to do something which you did not anticipate. The first decision box illustrates this problem. Someone may inadvertently specify an X matrix with more variables than observations (more columns than rows). If no test is made for this mistake, the com-

Table 13.1 List of Mathematical Expressions and Tentative APL Statements

Definition	Mathematical Expression	APL Expression
Moment matrix of X	$X'X$	MM←(⍉X)+.×X
Covariance matrix of X	$X'X - \bar{X}'\bar{X}$	CM←MM-(XB∘.×XB←+⌿X)÷(ρX)
Correlation matrix of X	$D^{-1/2}(X'X - \bar{X}'\bar{X})D^{-1/2}$	CRM←D+.×CM+.×D
(D is the diagonal matrix obtained from Cov(X))	$\text{Diag}(X'X - \bar{X}'\bar{X})$	←⌹(((ιK)∘.=ιK←(ρX)[2]) ×CM)*0.5
Regression coefficient estimators:	$\hat{B} = (X'X)^{-1}X'Y$ $\hat{Y} = X\hat{B}$ $\hat{U} = Y - \hat{Y}$	BE←Y⌹X YH←X+.×BE UH←Y-YH
Total Sum of Squares (TSS)	$\Sigma(Y - \bar{Y})^2$	TSS←+/(Y-YB)* 2 YB←(+/Y÷N)
Error Sum of Squares (ESS)	$\Sigma(Y - \hat{Y})^2 = \Sigma\hat{U}^2$	ESS←+/UH*2
$\hat{\text{Var}}$ (U)	$\Sigma\hat{U}^2/(N - K - 1)$	VARU←ESS÷(N-K+1)
R^2	$1 - ESS/TSS$	RSQ←1-ESS÷TSS
Regression SS (RSS)	$\Sigma(\hat{Y} - \bar{Y})^2$	
If done without constant term:		RSS←+/(YH-YB)*2
$F = (RSS/ESS)/((N - K - 1)/K)$		F←(RSS÷ESS)×((N-K+1)÷K)
COV matrix for $\hat{B}$, where $Z = (i, x)$	$\hat{\text{var}}$ $(U)(Z'Z)^{-1}$	COVB←VARU×⌹((⍉Z)+.×Z)
Standard deviations of elements of B given by square roots of diagonal elements of covariance matrix for $\hat{B}$, $S_{\hat{B}}$	$S_{\hat{B}} = (\text{Diag Cov}(\hat{B}))$	SBE←(1 1⍉COVB)*0.5
Student "t" ratios	$\hat{B}_i/S_{Bi}$, $i = 1, 2, \ldots, k$	TRATIO←BE÷SBE

puter will blithely proceed to the calculation of the regression coefficients and then give a *DOMAIN ERROR* and specify the line in the program at which the error occurred. This is all very well, but you may want the program which contains the regression routine to continue in any case and not stop. If so, a diagnostic test will get you around the difficulty before you waste valuable computer time. In addition, you can specify a warning variable to indicate when the regression routine has not, in fact, been executed; this is called "setting a flag."

The second idea is that sometimes we want to carry out the F test on the mean as well as on the statistical significance of the regression coefficients for the regressor matrix, so we have allowed for two variants of the test:

$$RRS_1 = \Sigma(\hat{Y} - \bar{Y})^2 \quad \text{and} \quad RSS_2 = \Sigma(\hat{Y})^2$$

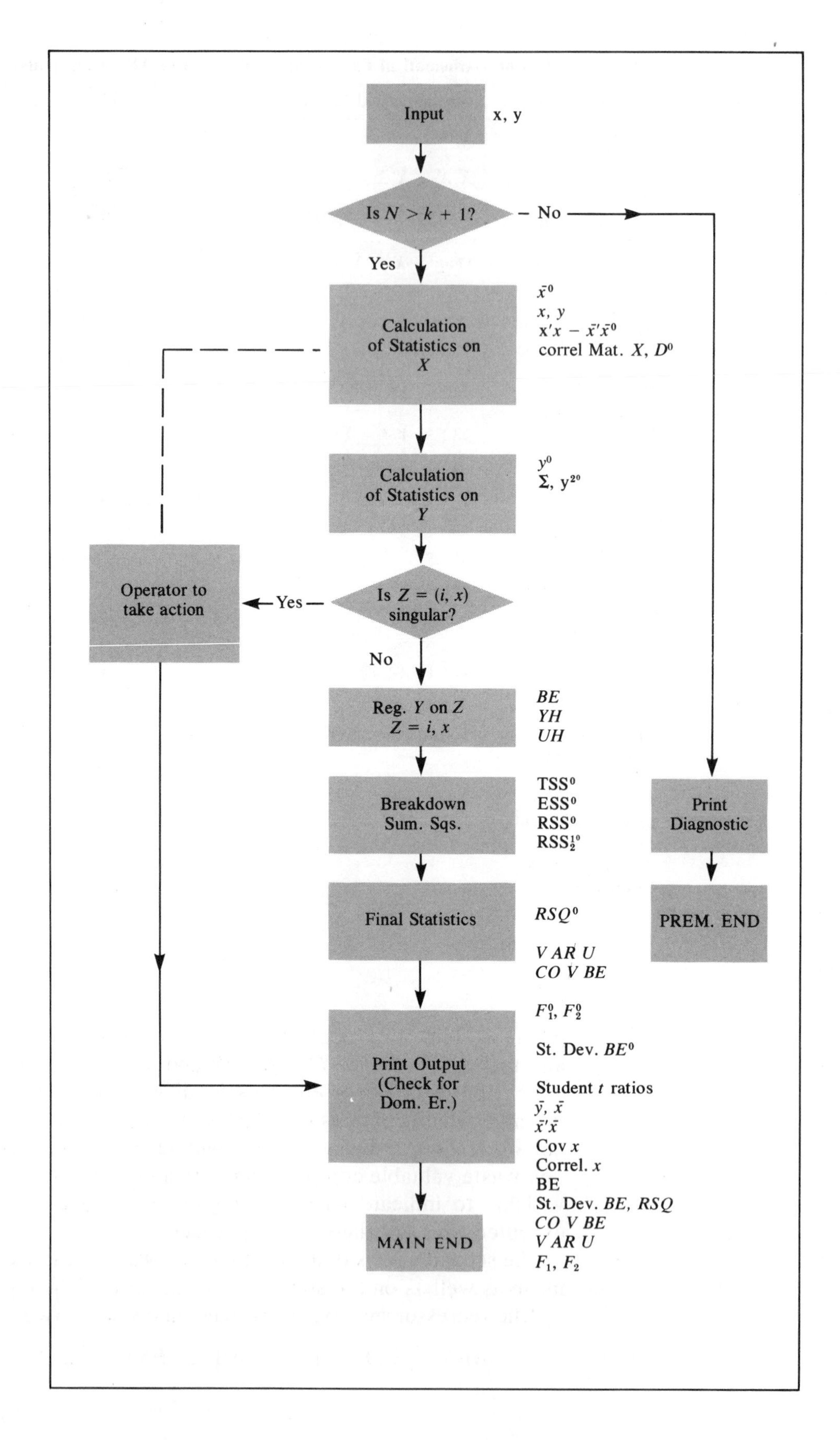

Figure 13.1
Flow chart for regression routine.

Sample Main Sheet

```
      ∇ Y MREGRESS X;MS;D;SS
[1]     ⍝ THIS PROGRAM ASSUMES Y TO BE AN ARRAY OF DIMEN.
        (N), X TO BE AN (N×K) ARRAY
[2]     ⍝ OF RANK K. A DIAGNOSTIC IS PRINTED IF N≤K OR IF
        RANK (X) <K.
[3]     ⍝ THE CONSTANT TERM IS ADDED BY THE ROUTINE.
[4]     NP← ⍴X
[5]     PREMEND1×⍳(NP[1]≤NP[2]+1)
[6]     CM←(MM←(⍉ X)+.×X)-(MS∘.×MS←+⌿X)÷NP[1]
[7]     CRM←D+.×CM+.×D←⌹(((⍳K)∘.=⍳K)×CM)*0.5
[4.2] K←NP[2]
[8]     MS←((+/Y),MS)÷NP[1]
[9]     SS[2]←(SS[1]←+/Y*2)-NP[1]×MS[1]*2
[4.5] SS←5 ⍴0
[10]    MAIN:UH←(Y-(YH←X+.×(BE←Y⌹(X←(1,X)))))
[3.1] ⍝**** IF DOMAIN ERROR OCCURS ON LINE [7] OR [MAIN]
        X IS SINGULAR, ROUTINE WILL BE SUSPENDED
[3.2] ⍝OPERATOR SHOULD KEY IN: →PREMEND
        IN ORDER TO COMPLETE ROUTINE. ****
[11]    SS[4]←(SS[3]←+/YH*2)-NP[1]×MS[1]*2
[12]    SS[5]←+/UH*2
[13]    RSQ←1-SS[5]÷SS[2]
[14]    VARU←SS[5]÷(-/NP)-1
[15]    F1←SS[4]÷SS[5]×K÷(-/NP)-1
[16]    F2←SS[3]÷SS[5]×(K+1)÷(-/NP)-1
[17]    COVBE←VARU×⌹(⍉X)+.×X
[18]    STDBE←(1 1 ⍉COVBE)*0.5
[19]    'THE RAW MOMENT MATRIX OF REGRESSORS IS:'
[20]    MM
[21]    'THE COVARIANCE MATRIX OF REGRESSORS, UNADJUSTED
        FOR THE NUMBER OF OBSERVATIONS, IS:'
```

```
[22]  CM
[23]    MAINEND:  'THE CORRELATION MATRIX OF REGRESSORS IS:'
[24]    CRM
[25]    'THE VECTOR OF MEANS; Y,X1,X1, . . .IS:'
[18.5]  TRATIO←BE÷STDBE
[26]    MS
[22.1]  →MAINEND×⍳FLAG
[22.3]  'ROUTINE PREMATURELY ENDED DUE TO SINGULARITY OF
        X MATRIX'
[22.4]  →0
[3.4]   FLAG←1
[18.1]  →CONT
[18.2]  PREMEND:FLAG←0
[18.3][19]  CONT:  'THE RAW MOMENT MATRIX OF REGRESSORS IS:'
[30]    'THE REGRESSION COEFFICIENT ESTIMATES ARE IN ORDER
        CONST., X1,X2 . . . :'
[31]    BE
[32]    'THE CORRESPONDING STANDARD ERRORS ARE:'
[33]    STDBE
[34]    'THE REG. COEF. EST. COVARIANCE MATRIX IS:'
[35]    COVBE
[36]    'RSQ IS: ';RSQ;' VAR. ERROR TERM IS: ' ;VARU
[37]    'THE F STATISTIC INCLUDING THE CONSTANT TERM IS:'
[38]    F2;' WITH ';(K+1),(¯1+-/NP); 'DEGREES OF FREEDOM. '
[39]    'THE F STATISTIC NOT INCLUDING CONSTANT TERM IS:'
[40]    F1; 'WITH ';K,(¯1+-/NP); ' DEGREES OF FREEDOM: '
[41]    '     THIS ENDS THE OUTPUT FROM MREGRESS. '
[42]    →0
[43]    PREMEND 1:'NO. OF OBS. (N) IS TOO FEW RELATIVE TO NO.
        OF REGRESSORS (K).'
[44]    'ROUTINE TERMINATED'
```

```
[33.5]  'THE CORRESPONDING T RATIOS ARE:'
[33.7]  TRATIO
[45]    ∇
```

Sample Subsidiary Sheet

Used for Preliminary Working Out of APL Statements

Reline [4]:
```
N←ρX
→PREMEND×ι(N[1]≤N[2]+1)
```

Reline [6]:
```
MM←(⍉X)+.×X
CM←MM-(MS∘.×MS←+⌿X)÷N[1]
```

Reline [7]:
```
CRM←D+.×CM+.×D←⌹(((ιK)∘.=ιK←N[2])×CM)*0.5
```

Reline [8]:
```
MS←((+/Y)÷N[1],MS÷N[1])
```

Reline [9]:
```
TSS2←(TSS←+/Y*2)-N×MS[1]*2
```

Reline [10]:
```
YH←X+.×(BE←Y⌹(X←(1,X)))
```

At this point you could get a *DOMAIN ERROR* and a suspended routine if the augmented *X* matrix is singular. To get out of this and continue operation will require operator intervention from the terminal; a flag must be set to bypass the remaining analysis and to print only what has been done so far. Operator action is indicated in the comment lines labelled [3.1] and [3.2]. You will notice that we have *just realized* that we could also get a domain error leading to a suspension in the current line [7]. (This is indicated on the flow chart by the dotted-dashed line which was added later.)

Reline [11]:
```
SS[4]←(SS[3]←+/YH*2)-NP[1]×MS[1]*2
```

Reline [17]:
```
COVBE←VARU×⌹(⍉X)+.×X
```

Reline [18]:
```
STDBE←(1 1⍉COVBE)*0.5
```

Some of the early calculations not affected by the domain error can be printed, but we will need to bypass the remaining items if the routine has a domain error. The inserts at lines [3.4] and at [18.1;2;3] do this.

With the flow chart and our previously developed "one-liners" in hand, let us proceed to write out our regression routine. As you do it, start off with a large clean piece of paper, work in pencil, and leave yourself lots of room for making corrections. Do the more complicated one-liners first on a separate piece of paper, then fit them into the main routine. When you have finished, rewrite the routine carefully, checking it as you go. Review it in light of our comments above, then (and only then), consider keying into the computer. Here is a typical sample of trying to write a routine—warts and all! Before you read the following paragraphs, glance over the sample program listed on pages 251–253, the detailed comments on pages 246–247, and the flow chart in Figure 13.1.

Some comments are in order. While all the APL operations used in this program are familiar to you by now, there are a few items which should be emphasized. In line [9] we decided we would put the various sums of squares into an array. But executing `SS[1]←+/Y*2` would give a value error if *SS* had not already been declared a list! That is why we had to insert line [4.5], i.e., this line is to go between lines [4.2] and [5]. Line [4.5], `SS←5 ρ 0`, sets up *SS* as an array of zeros. The statements in line [9] put sums of squares into various elements of the list *SS*.

The inserted lines [4.5], [3.1], and [3.2] illustrate two things. First, we promised that we would show how writing a routine is done, including mistakes! Quite frankly, we forgot all about the need to define *SS* as an array before reaching line [9] until we, in fact, wrote line [9]. Likewise, the inserted statements [3.1] and [3.2] were afterthoughts—in short, we goofed. The second item to note is that the routine can be keyed-in *just as it stands,* and on being displayed everything will be rearranged into proper sequence by the computer.

Finally, as you look between the flow chart, the preliminary statement, and the sample main sheet version, you will see that we changed our mind at times; we hope each change was for the better. If nothing else, this shows that the flow chart is a guide to aid you, not an inflexible route from which you dare not deviate.

The routine is not quite finished, since we need to add a series of comment statements to guide any reader (including ourselves in a month's time). For example,

```
[4.8]  ⍝  CHECK NO. OBS.  (N) > NO. VARS (K)
[5.1]  ⍝  THIS SECTION CALCULATES RAW MOMENT MATRIX, MOM. MAT.,
          AND CORREL. MAT. FOR REG. MAT. X.
```

and so on.

A Regression Routine

The last steps are to reread the routine carefully to check for errors of both omission and commission, key it into the computer, check that the keying-in is correct, and then run some tests on its operation under all the situations it is meant to encounter. For example, (i) run a regression for which the results are already known; (ii) try to enter a regression with fewer observations than variables; and (iii) try one with a singular *X* matrix.

Let's try the routine. Begin by displaying the rearranged routine and the *X* and *Y* arrays to be used for the test.

```
          ∇MREGRESS[⎕]∇

        ∇ Y MREGRESS X;MS;D;SS
[1]       ⍝THIS PROGRAM ASSUMES Y TO BE AN ARRAY OF DIMEN.  (N),
[2]       ⍝X TO BE AN (N×K) ARRAY OF RANK K. A DIAGNOSTIC IS
[3]       ⍝PRINTED IF N≤K OR IF RANK X < K
```

```
[4]     ⍝THE CONSTANT TERM IS ADDED BY THE ROUTINE.
[5]     ⍝*****  IF DOMAIN ERROR OCCURS ON LINE 14 OR [MAIN] X IS
          SINGULAR,
[6]     ⍝ROUTINE WILL BE SUSPENDED.  OPERATOR SHOULD KEY IN: →PREMEND
[7]     ⍝IN ORDER TO COMPLETE ROUTINE.
[8]      FLAG←1
[9]      NP←ρX
[10]     K←NP[2]
[11]     SS←5 ρ0
[12]     →PREMEND1×ι(NP[1]≤NP[2]+1)
[13]     CM←(MM←(⍉X)+.×X)-(MS∘.×MS←+⌿X)÷NP[1]
[14]     CRM←D+.×CM+.×D←⌹(((ιK)∘.=ιK)×CM)*0.5
[15]     MS←((+/Y),MS)÷NP[1]
[16]     SS[2]←(SS[1]←+/Y*2)-NP[1]×MS[1]*2
[17]     MAIN:UH←(Y-(YH←X+.×(BE←Y⌹(X←(1,X)))))
[18]     SS[4]←(SS[3]←+/YH*2)-NP[1]×MS[1]*2
[19]    SS[5]←+/UH*2
[20]    RSQ←1-SS[5]÷SS[2]
[21]    VARU←SS[5]÷(-/NP)-1
[22]    F1←SS[4]÷SS[5]×K÷(-/NP)-1
[23]    F2←SS[3]÷SS[5]×(K+1)÷(-/NP)-1
[24]    COVBE←VARU×⌹(⍉X)+.×X
[25]    STDBE←(1 1 ⍉COVBE)*0.5
[26]     TRATIO←BE÷STDBE
[27]    →CONT
[28]    PREMEND:FLAG←0
[29]    CONT:'THE RAW MOMENT MATRIX OF REGRESSORS IS:'
[30]    MM
[31]    'THE COVARIANCE MATRIX OF REGRESSORS IS:'
[32]    CM
[33]    →MAINEND×ιFLAG
[34]    'ROUTINE PREMATURELY ENDED DUE TO SINGULARITY OF X MATRIX'
```

```
[35]   →0
[36]   MAINEND:'THE CORRELATION MATRIX OF REGRESSORS IS:'
[37]   CRM
[38]   'THE VECTORS OF MEANS,Y, X1, X2, X3, ......IS:'
[39]   MS
[40]   'THE REGRESSION COEFFICIENT ESTIMATES ARE IN ORDER CONST.,
       X1,X2,.....:'
[41]   BE
[42]   'THE CORRESPONDING STANDARD ERRORS ARE.'
[43]   STDBE
[44]   'THE CORRESPONDING T RATIOS ARE:'
[45]   TRATIO
[46]   'THE COVARIANCE MATRIX OF THE REGRESSION COEF. IS.'
[47]   COVBE
[48]   'RSQ IS:';RSQ;'VAR OF ERROR TERM IS:';VARU
[49]   'THE F STATISTIC INCLUDING THE CONSTANT TERM IS:'
[50]   F2;'WITH';(K+1),(¯1+-/NP);'DEGREES OF FREEDOM.'
[51]   'THE F STATISTICS NOT INCLUDING CONSTANT TERM IS:'
[52]   F1;'WITH ';K,(¯1+-/NP);'DEGREES OF FREEDOM.'
[53]   'THIS ENDS THE OUTPUT FROM MREGRESS.'
[54]   →0
[55]   PREMEND1:'NO. OF OBS  (N) IS TOO FEW RELATIVE TO NO! OF
       REGRESSORS  (K) .'
[56]   'ROUTINE TERMINATED'
       ∇
      Y
100 106 107 120 110 116 123 133 137
      X
100 100
104  99
106 110
111 126
```

```
111 113
115 103
120 102
124 103
126  98
```

[These data are from Johnston, *Econometric Methods,* p. 147.]

```
      Y MREGRESS X
THE RAW MOMENT MATRIX OF REGRESSORS IS:
 115571 107690
 107690 101772
THE COVARIANCE MATRIX OF REGRESSORS IS:
 650     ¯112
¯112      648
THE CORRELATION MATRIX OF REGRESSORS IS:
 1       ¯0.17257
¯0.17257  1
THE VECTORS OF MEANS,Y, X1, X2, X3, ......IS:
116.89 113 106
THE REGRESSION COEFFICIENT ESTIMATES ARE IN ORDER CONST.,X1,X2,...:
¯49.341 1.3642 0.11388
THE CORRESPONDING STANDARD ERRORS ARE.
24.061 0.14315 0.14337
THE CORRESPONDING T RATIOS ARE:
¯2.0507 9.5299 0.79429
THE COVARIANCE MATRIX OF THE REGRESSION COEF. IS.
 5.7893E2     ¯2.6911E0     ¯2.5792E0
¯2.6911E0      2.0493E¯2     3.5420E¯3
¯2.5792E0      3.5420E¯3     2.0556E¯2
RSQ IS:0.9385VAR OF ERROR TERM IS:12.924
THE F STATISTIC INCLUDING THE CONSTANT TERM IS:
3202.2WITH3 6DEGREES OF FREEDOM.
THE F STATISTICS NOT INCLUDING CONSTANT TERM IS:
```

```
45.782WITH 2 6DEGREES OF FREEDOM.

THIS ENDS THE OUTPUT FROM MREGRESS.
```

Now let's try the oddball cases. Consider:

```
      Q←1 2 3
      XX←3 4 ρ1 2 3 4 5 6 7 8 9 10 11 12
      Q MREGRESS XX
NO. OF OBS.(N) IS TOO FEW RELATIVE TO NO. OF REGRESSORS(K).
ROUTINE TERMINATED.
      Q←ι5
      XX←5 3 ρ1 2 3 1 5 6 1 8 9 1 10 11 1 12 13
      Q MREGRESS XX
DOMAIN ERROR
MREGRESS[14] CRM←D+.×CM+.×D←⌹(((ιK)∘.=ιK)×CM)*0.5
      →PREMEND
THE RAW MOMENT MATRIX OF REGRESSORS IS:
  5   37   42
 37  337  374
 42  374  416
THE COVARIANCE MATRIX OF REGRESSORS IS:
      0              0              0
      0           63.2           63.2
      0           63.2           63.2
ROUTINE PREMATURELY ENDED DUE TO SINGULARITY OF X MATRIX.
```

13.5 Bells and Whistles Section

In this section we will discuss some ways in which this simple multiple linear regression model can be applied to a wider variety of circumstances than might at first appear to be the case.

Transformed Variables

The first way in which the linear model can be extended is to broaden the concept of the regressor matrix used in the regression routine. Let us suppose that you have a number of arrays called, say, $X1$, $X2$, $X3$, and so on. The arrays might represent time series of interest rates, net national product, consumption expenditures, population, consumer price index, money supply, etc. We might call these the raw data. In the regression model we have derived from other considerations, we may want to use not the raw data but functions of them, such as

$$\ln(x_{ij}),\ e^{x_{ij}},\ x_{ij}^2,\ x_{ij}/x_{kj},\ (x_{ij} + x_{kj})/x_{ij},\ (x_{ij}x_{kj})^2$$

and so on, where x_{ij}, x_{kj} represent the jth observation on variables x_i and x_k. These transformations are easily handled in APL. For example:

```
(,Y÷10*5) MREGRESS  ⍉(3, NOBS) ρ(,X1÷X2),(,⍟X3), (,X3+X4)÷X5
```

Transformed Variables

enables one to transform variables by arrays and set up the regression without having to define a new regressand vector or a new regressor matrix; and all of this is done in one statement. *NOBS* is the value of the number of observations.

Heteroskedastic Models

Heteroskedastic Models

One important special case of these ideas is the concept of weighted regression which frequently arises in heteroskedastic models. These are models wherein the covariance matrix of the disturbance term is a diagonal matrix with *unequal* diagonal elements. Thus, suppose that the variance of the ith disturbance term, $i = 1, 2, \ldots, N$, there being N observations, is $\sigma^2 z_i$, where σ^2 is an unknown scalar and z_i is a *known* constant. Then, if the ith observations on the regressand and the regressors are divided by $\sqrt{z_i}$, the transformed model meets the usual multiple linear regression model assumption that the covariance matrix of the disturbance term array is $\sigma^2 I$, where I is the identity matrix. (For further details about such models see Kmenta in the bibliography.) This transformation is performed without difficulty in APL.

Lagged Variables

The next problem is a little trickier. Suppose that we have three arrays labelled Y, $X1$, and $X2$ representing time series on three variables of interest. Now, as frequently is the case in econometric analysis, you want to carry out your analysis not just with the original three arrays, but with arrays created by *lagging* the original arrays. An associated problem is that the lagging process will reduce the usable length of our arrays. Let's consider a lag of length M; M can be 1 or 2 or any positive integer less than $(N - K - 1)$, where N is the number of observations (or length of the arrays) and K is the number of regressors to be used in the regression analysis.

Lagged Variables

With lagging, the regressors will include not only $X1$ and $X2$ (the original regressors), but also the lagged values of Y, $X1$, and $X2$. With lags of length

M, and N the length of the original array, the variable length of the new array is only $(N - M)$. Let's call the lagged variables YL, $X1L$, and $X2L$. Consider, therefore, a lag of length M:

```
YL ← Y[P ← ιN-M]

X1L ← X1[P]

X2L ← X2[P]

K←((YL←K/Y),(X1L←K/X1),(X2L←(K←(Mρ0),(N-M)ρ1)/X2))[ι0]
```

Now ask for YL, $X1L$ and $X2L$. The reader might recall that here we have used the dyadic form of /, called compression. Essentially, we have dropped the first M elements of the arrays on the right.

We now have six arrays of length $(N - M)$, the former three of which—Y, $X1$, $X2$—contain observations from the $(M + 1)$st to the Nth, and the latter three—YL, $X1L$, $X2L$—contain the first $(N - M)$ observations from Y, $X1$, and $X2$, respectively. Just to check these ideas out, let's try it, using the Y array and X matrix defined earlier. Let M be 2. Thus we key in:

```
      Y
100 106 107 120 110 116 123 133 137
      X
1  100  100
1  104   99
1  106  110
1  111  126
1  111  113
1  115  103
1  120  102
1  124  103
1  126   98
      ρX
9 3
      M←2
      YL←Y[P←ιN-M]
      XL←X[P ;1 2 3]
      Y←((M ρ0),(N-M)ρ1)/Y
      X←(( Mρ0),(N-M)ρ1)⌿X
```

```
        Y
107 120 110 116 123 133 137
        X
1  106  110
1  111  126
1  111  113
1  115  103
1  120  102
1  124  103
1  126   98
```

A comparison of the new arrays *Y* and *YL* and the matrices *X* and *XL* will show that we have accomplished our objective. Examine the use of parentheses very carefully in the expression `((Mρ0) , (N-M)ρ1)`.

Durbin-Watson Statistic

Durbin-Watson Statistic

Now consider a useful test statistic in linear regressions involving time series—the Durbin-Watson statistic (see, for example, Kmenta, pages 295–97). From *MREGRESS* we get a globally defined array called *UH*, which contains the forecast error terms. Mathematically, the Durbin-Watson statistic is defined by

$$d = \Sigma_{t=2}^{T}(e_t - e_{t-1})^2 / \Sigma_1^T e_t^2$$

where $e_t = (y_t - \hat{y}_t)$. An obvious APL solution is

```
A ←+/UH*2
D ←(÷A)×+/(((0,(T-1) ρ1)/UH)-((T-1) ρ1,0)/UH)*2
```

Try it on the *UH* generated by *MREGRESS:*

```
     ∇D←DURWAT X
[1]  T1←(ρX)-1
[2]  A←+/X*2
[3]  D←(÷A)×+/(((0,T1 ρ1)/X)-((T1 ρ1),0)/X)*2
[4]  'THE DURBIN-WATSON TEST STATISTIC WITH ';T1;' DEGREES
     OF FREEDOM IS:'D
     ∇
     Y MREGRESS X
```

The computer output for *MREGRESS* has been suppressed to save space (see page 257).

```
      DURWAT UH
THE DURBIN-WATSON TEST STATISTIC WITH 8 DEGREES OF FREEDOM IS:
1.635297783
```

Summary

For a matrix X of N observations and K variables, the following statistics can be defined:

Moment Matrix	$X'X$
Covariance Matrix	$X'X - \bar{X}'\bar{X}$
Correlation Matrix	$D^{-1/2}(X'X - \bar{X}'\bar{X})D^{-1/2}$, where D contains the diagonal elements of $(X'X - \bar{X}'\bar{X})$

[See page 241 for the APL expressions.]

Sample Correlation Coefficients:

$$r_{ij} = \mathrm{Cov}(x_i, x_j)/\sqrt{\mathrm{Var}(x_i)\,\mathrm{Var}(x_j)}$$

where $\mathrm{Cov}(x_i, x_j)$, $\mathrm{Var}(x_i)$, and $\mathrm{Var}(x_j)$ are sample values.

Sample Partial Correlation Coefficients:

$$r_{ij\cdot k} = (r_{ij} - r_{ik}r_{kj})/\sqrt{(1 - r_{ik}^2)(1 - r_{jk}^2)}$$

For regression statistics, see Table 13.1. The corresponding APL statements are contained in *MREGRESS* on page 254.

Durbin-Watson Test Statistics:

$$d = \Sigma_{t=2}^{T}(e_t - e_{t-1})^2/\Sigma_1^T e_t^2$$

where $e_t = (Y_t - \hat{Y}_t)$.

The APL routine is listed on page 261.

Exercises

APL Practice

1. Suppose that you have an array ZZ. How would you instruct the computer to form a diagonal matrix D with diagonal elements those of ZZ? Specify an array ZZ and check your routine.

2. Write a simple routine called *CORRMAT*, whose output will consist of the correlation matrix of the regressors with their respective names, i.e.:

```
THE CORRELATION MATRIX OF REGRESSORS

NAME       JJ        KK        LL       . . .

JJ          *         *         *       . . .

KK          *         *         *       . . .

LL          *         *         *       . . .

.           .         .         .

.           .         .         .

.           .         .         .
```

Have the program ask you for the names of the variables first and then print the correlation matrix.

3. Write a program that will calculate the partial correlation coefficients of a set of regressors X.

4. Write a simple routine that will give you the regression coefficients in the following format:

```
                   ESTIMATED REGRESSION COEFFICIENTS

                          STANDARD ERROR     T-STATISTIC

CONSTANT      *                *                  *

COEF. OF X1 *                  *                  *

COEF. OF X2 *                  *                  *
```

Furthermore, if one of the t-statistics is smaller than 1.96, print a message to state which coefficient is not significantly different than zero at the 5% significance level.

5. Another way to calculate the Durbin-Watson statistic is given by the formula

$$d = \frac{\hat{u}'A\hat{u}}{\hat{u}'\hat{u}}$$

where $\hat{u}$ is a vector of estimated error terms and

$$A = \begin{bmatrix} 1 & -1 & 0 & 0 & . & & & & \\ -1 & 2 & -1 & 0 & . & & 0 & & \\ 0 & -1 & 2 & -1 & . & & & & \\ . & . & . & . & . & . & . & . & . \\ & & & & . & -1 & 2 & -1 & 0 \\ & 0 & & & . & 0 & -1 & 2 & -1 \\ & & & & . & 0 & 0 & -1 & 1 \end{bmatrix}$$

Write a short routine to calculate the Durbin-Watson statistic using this matrix formulation.

For the following list of residuals calculated as $UH = 1, -1, 2, 3, -4, -5, 1, 3$, find the Durbin-Watson statistic using the above formula and the `DURWAT H` function (page 261). Do you get the same result? If not, why not?

Statistical Applications

The questions in this section utilize the Data Set Watt in Appendix E.

1. Consider the following one equation production model

$$Y = b_0 + b_1X_1 + b_2X_2 + U$$

where Y is production in KWHR of 15 electricity-generating plants, X_1 is the input in BTU's per firm, and

$$X_2 = \begin{cases} 1 & \text{if plant's number is odd} \\ 2 & \text{if plant's number is even} \end{cases}$$

Test the following pair of hypotheses using only the second year data (1967)

a. H_0: $b_0 = 0, b_2 = 0$ against H_1: $b_0 = 0, b_2 \neq 0$

and compare to:

b. H_0: $b_2 = 0$, b_0 unrestricted against H_1: $b_2 \neq 0$, b_0 unrestricted

(Hint: The column of 1's should be used.)

c. Does your answer regarding the importance of X_2 in the model differ between tests (a) and (b)?

2. Suppose that the manager of the second power plant knows that one of the following two models is the true model that explains the factor demand for energy. The two models are

model A $\quad Y_t = a_0 + a_1X_t + a_2P_t + U_{1t}$

model B $\quad Y_t = bX_tP_t + U_{2t}$

where Y_t is input in BTU's for year t, P_t is the price per BTU paid by the second plant for year t, X_t is the production in KWHR for year t, and the standard assumption that the distributions of U_{it}, $i = 1, 2$ are given by:

$$U_{1t} \sim N(0, \sigma_1^2)$$

$$U_{2t} \sim N(0, \sigma_2^2)$$

Show that model A is the true model. (Hint: combine the two models and test the coefficient on the X_tP_t term.)

3. The purchasing department of the *sixth* plant has the following model to estimate the factor demand relationship:

$$Y_t = a_0 + a_1X_t + a_2P_t + a_3D + U_t$$

where Y_t is the quantity of energy utilized in BTU's for year t, P_t is the price per BTU paid by the plant in year t, X_t is the output in KWHR for year t, and D is a dummy variable which takes on the value one if summer temperature exceeds 90°F, and the value zero if the temperature is less than 90°F. The summer temperatures are:

Year	66	67	68	69	70	71	72	73	74	75
Temp.	100	92	92	94	99	92	105	92	97	87

(a) Find the estimates of the coefficients and the estimates of their covariance matrix.

(b) Drop the 1975 observation. Notice that the matrix of regressors is now singular. However, one can obtain estimates of these parameters a_1, a_2, and a^*, where $a^* = a_0 + a_3$; that is, drop the variable D. What are the estimates of a_1, a_2, and a^*? Determine the covariance matrix of a_1, a_2, and a^*. Compare estimates of a_1 and a_2 with your answers to (a). Use the appropriate column of the data set WATT in Appendix E.

4. Suppose the Department of Energy wanted to know the production relationship between BTU and KWHR. A model that they might use employs averages for the variables of the industry. The model could be

$$\bar{Y}_t = a_0 + a_1\bar{X}_t + a_2\bar{P}_t + U_t$$

where $\bar{Y}_t$ is the average output (in KWHR) of the plants in year t, $\bar{X}_t$ is the average input in BTU's to the plants in year t, and $\bar{P}_t$ is the average price per BTU paid by the plants in year t. Here $t = 1, \ldots, 10$.
Estimate the coefficients of this model.
Estimate the covariance matrix of the coefficients.
Use data set WATT of Appendix E.

5. The manager of the fourth plant conceived a more sophisticated model for his input demand. He believed that his input demand for fuel is explained by a partial adjustment model. The model is

$$\text{(i)} \qquad Y_t^* = b_0 + b_1P_t + b_2X_t$$

$$\text{(ii)} \qquad Y_t - Y_{t-1} = \gamma(Y_t^* - Y_{t-1}) + U_t$$

where Y_t is BTU for year t, P_t is the prices for year t, X_t is KWHR for year t, and Y^* is the unobserved desired Y_t. As before, $t = 1, \ldots, 10$.
Using data set WATT of Appendix E:

(a) Show that the input demand equation is equivalent to

$$Y_t = b_0\gamma + b_2\gamma X_t + (1 - \gamma)Y_{t-1} + b_1\gamma P_t + U_t$$

(b) Run an OLS and obtain the estimates of the coefficients. (Assume that $Y_{1965} = 700 \times 10^{12}$ BTU's).

6. In the model of problem 5, notice that the first equation is deterministic; i.e., it has no error term. Suppose that the manager believed instead that the following model was correct:

(i) $$Y_t^* = b_0 + b_1P_t + b_2X_t + U_t$$

(ii) $$Y_t - Y_{t-1} = \gamma(Y_t^* - Y_{t-1})$$

(a) Solve for Y_t as you did in part (a) of problem 5.

(b) Estimate the coefficients using the least squares procedure discussed in this chapter.

7. Suppose that you hypothesize that current production of KWHR depends on current prices of the input (BTU) and the previous year's prices of the input as well. Your model now becomes $Y_t = a_0 + a_1P_t + a_2P_{t-1} + U_t$ for the first plant. Test the hypothesis that your model is the correct model against the alternative, that KWHR depends on BTU's consumed and current prices, written as $Y_t = a_0 + a_1X_t + a_2P_t + U_t$ is the correct model, using data set WATT of Appendix E.

8. In this application you will extend the idea of adding lagged prices to the demand function as was done in the previous exercise. Suppose that you hypothesize that it takes two to three years to adjust output to price changes in inputs. Formulate your model and test the hypothesis that the adjustment period is three years, using data from the first plant. The two models can be written as:

(i) $Y_t = a_0 + a_1P_t + a_2P_{t-1} + a_3P_{t-2} + a_4P_{t-3} + U_{1t}$
(ii) $Y_t = b_0 + b_1X_t + b_2P_t + U_{2t}$

where the variables are defined in question 2.

9. Consider the following model:

$$Y_{it} = a + b_{it}X_{it} + U_{it} \qquad i = 1, \ldots, 15$$

where Y_{it} is the 10 by 1 column vector of KWHR of firm i over the 10 years, X_{it} is the 10 by 1 column vector of BTU's used by firm i, and over the 10 years b_{it} is the coefficient relating input to output. Test the following two hypotheses:

(a) The coefficient b_{it} is the same for all of the 15 plants, although they may differ over time, i.e., $b_{it} = b_t$ for all i.

(b) The relationship between X and Y remains constant over time, but it might differ across firms, i.e., $b_{it} = b_i$ for all t.

10. If you have already found that the functional form of the relationship between prices of BTU's and BTU's demanded by the whole industry is

$$\text{BTU} = a_0 P^{a_1} e^{u}$$

(a) Estimate a_0 and a_1.
(b) Find the price elasticity of demand for inputs for the entire industry.

14

Other Simple Regression Equation Estimators

You will not study econometrics very long before you discover that the linear multiple regression model we discussed at length in the previous chapter cannot handle all eventualities, and so new estimators have to be created. This chapter will discuss a number of these.

14.1 Simultaneous Equation Models

Simultaneous Equation Models

One of the most important extensions of the linear regression model is based upon the recognition that in most situations in economics we cannot consider a single regression equation in isolation, but instead must treat it as merely one equation within an interdependent system. In short, in this chapter we begin by introducing the notion of *simultaneous equation models*. For an excellent introduction to this fascinating area of econometrics see the books by Kmenta or Johnston (listed in the bibliography).

The most usual simple example is to consider the market for lemons (the fruit, not inferior cars or appliances!). If you have studied only a little economics you will be aware that the price and quantity of lemons exchanged in the market are the result of the *interaction* between two behavioral relationships, one determining producers' supply responses and one determining consumers' demand responses to market prices.

Now the statistical problem we face if we want to use observed data to estimate either response, or even both responses, in that we cannot blithely run a regression of, say, quantity traded, on price and other variables affecting demand and expect to get useful estimates of the parameters of the model. Besides, we might immediately wonder whether we should really be regressing price on quantity and other variables or quantity on price.

This is not the place for us to settle these issues since many books, indeed several very good books, have been written on the subject. We will proceed under the assumption that you have read or are reading a textbook on estimation in simultaneous equation systems.

The *estimation* problem in a simultaneous equation system arises from the distinction between endogenous and exogenous variables. So far in linear regression the distinction has been straightforward—the dependent variable is the endogenous variable and the regressors on the right-hand side of the equation are the exogenous variables. Endogenous variables are determined by the set of equations being considered and exogenous variables, while determined outside the system, affect the endogenous variables through the equations. The problem in a simultaneous equation system is that we have *more* than *one* endogenous variable in *each* equation! The regression techniques that were suitable for the single endogenous variable are not suitable for the many endogenous variables case. New techniques are needed. We now address ourselves to one of these cases.

14.2 Two-Stage Least Squares

2SLS Regression Procedure

One method of estimation for a single equation that is frequently used with simultaneous equation systems is called two-stage least squares (2SLS). Here is an example of a simplified model for food consumption from Kmenta, pages 563–656.

$$Q_t = a_1 + a_2P_t + a_3D_t + U_{dt} \qquad \text{(demand)}$$
$$Q_t = b_1 + b_2P_t + b_3F_t + b_4A_t + U_{st} \qquad \text{(supply)}$$

where Q_t is food consumption per capita, P_t is the ratio of food prices to general consumer prices, D_t is disposable income in constant prices, F_t is the ratio of the preceding year's prices received by farmers for products to general consumer prices, A_t is time in years, and U_{dt} and U_{st} are the disturbance terms. The data are presented on page 285.

In simultaneous equation systems we are usually interested in what are known as the structural coefficients, that is, the coefficients relating the effect of, say, income, on quantity demanded or price. The demand and supply equations shown above are structural equations. Alternatively, we can consider solving for each endogenous variable in terms of all the exogenous variables. These latter equations are called the "reduced form" equations because the prediction of each endogenous variable has been reduced to a linear function of exogenous variables only. Of course, the reduced form coefficients are in general complicated functions of the structural coefficients. For example, the reduced form equations for the above simple model are

$$Q_t = \gamma_0 + \gamma_1D_t + \gamma_2F_t + \gamma_3A_t + V_{1t}$$
$$P_t = \delta_0 + \delta_1D_t + \delta_2F_t + \delta_3A_t + V_{2t}$$

and the reduced form coefficients expressed as functions of the structural coefficients are

$$\gamma_0 = \frac{b_2 a_1 - a_2 b_1}{b_2 - a_2} \qquad \gamma_1 = \frac{b_2 a_3}{b_2 - a_2}$$

$$\gamma_2 = \frac{-a_2 b_3}{b_2 - a_2} \qquad \gamma_3 = \frac{-a_2 b_4}{b_2 - a_2}$$

$$\delta_0 = \frac{b_1 - a_1}{b_2 - a_2} \qquad \delta_1 = \frac{-a_3}{b_2 - a_2}$$

$$\delta_2 = \frac{b_3}{b_2 - a_2} \qquad \delta_3 = \frac{b_4}{b_2 - a_2}$$

$$V_{1t} = \frac{b_2 U_{dt} - a_2 U_{st}}{b_2 - a_2} \qquad V_{2t} = \frac{U_{st} - U_{dt}}{b_2 - a_2}$$

In this simple model Q_t and P_t are the endogenous variables and D_t, F_t, and A_t are the exogenous variables.

The 2SLS procedure starts with estimates of the coefficients of the reduced form equation, wherein each endogenous variable is a function of all the exogenous variables in the system. The second stage is to use predicted values of the endogenous variables as regressors in the regression equation, relating one endogenous variable to the others in the *structural* equation. These first round coefficients in the price equation are computed by

```
V←P ⌹1,D,F,A
```

The only thing new here is the catenation of *D, F, A* and a column of 1's, which is a simple extension of the way in which we used catenate earlier. Recall that the computer operates from right to left. You might type 1, D, F, A to see the matrix in full.

Next we compute the "estimated" or "fitted" values of the *P* list by multiplying each of the coefficients by the appropriate data value. For the first sample point we have

$$99.628 = 90.278\ (1) + 0.663\ (87.4) - 0.488\ (98.0) - 0.737(1)$$

In order to compute the vector of all fitted values we can use matrix multiply

```
(1,D,F,A) +.× V
```

and to store the result in a 20 by 1 vector we would write

```
VV← 20 1 ρ (1,D,F,A) +.× V
```

You may wish to type *VV* and display the 20 fitted values. This completes the first stage of the procedure.

The second stage is to estimate the structural equations using the fitted values of *P* (stored as *VV*) in place of the original values of *P*. For the demand equation, this is accomplished by

```
Q⌹1,VV,D

94.63333

0.24355

0.31399
```

These are the 2SLS estimators of the structural coefficients a_1, a_2, a_3.

The structural equation is

$$\hat{Q}_t = 94.63333 - 0.2436\ P_t + 0.3140\ D_t$$

Similarly, the supply equation is

```
Q⌹1, VV, F, A

49.53244

0.24007

0.25561

0.25292
```

These are the 2SLS estimates of the structural coefficients b_1, b_2, b_3, b_4.

or

$$\hat{Q}_t = 49.53244 + 0.24007\ P_t + 0.25561\ F_t + 0.25292\ A_t$$

You may have noticed that the supply equation is exactly identified; that is, the number of excluded exogenous variables from that equation is equal to the number of endogenous variables included in the regression as regressors. The excluded exogenous variable is D_t and the included endogenous variable is P_t.

In the exactly identified case we can derive estimates of the structural coefficients directly from the estimates of the coefficients of the reduced form equation. This is usually an exercise in algebra, where the values of the structural coefficients to be estimated are obtained by solving the equations relating the structural and reduced form coefficients in the exactly identified equation. To accomplish this we can estimate the reduced form equations for P and Q.

```
Q⌹1,D,F,A

71.20355

0.15922

0.13834

0.07598
```

The estimates of the reduced form coefficients γ_0, . . . , γ_3.

```
P⌹1,D,F,A

90.26776

0.66321

-0.48845

-0.737039
```

The estimates of the reduced form coefficients δ_0, . . . , δ_3.

Algebraically, we have

$$\begin{aligned}\hat{Q} &= 71.20355 + 0.15922D + 0.13834F + 0.075978A \\ &= B_1 + B_2(90.26776 + 0.66321D - 0.48845F - 0.737039A) \\ &\quad + B_3F + B_4A\end{aligned}$$

After some algebraic manipulation we obtain

$$B_2 = \frac{0.15922}{0.66321} = 0.24007$$

$$B_3 = 0.13834 - (0.24007)(-0.48845) = 0.2556$$

etc., which are in fact our 2SLS results. The other coefficients are left as an exercise for the reader. It should be clear that even when the equations are exactly identified, 2SLS is a much simpler procedure than is indirect least squares. Of course, when the equation is overidentified (more excluded exogenous variables than included endogenous ones), you cannot use the indirect least-squares procedure, but you can use 2SLS.

As a final example, you may want to see what results you would have obtained if you had performed ordinary least squares (O.L.S.) on the structural equations. (Remember that these estimates are statistically inconsistent.)

The demand equation would be

```
      Q⌹1,P,D
                                              94.63333
99.89542                                       0.24355
-0.31630                                       0.31399
0.33464
```

←The O.L.S. estimates
The 2SLS estimates→

and the supply equation would be

```
      Q⌹1,P,F,A
                                              49.53244
58.27543                                       0.24007
0.16037                                        0.25561
0.24813                                        0.25292
0.24830
```

←The O.L.S. estimates
The 2SLS estimates→

The result of this section is that you can use APL to compute estimates that have desirable statistical properties. The procedures are not difficult. Even in the case of exactly identified equations, 2SLS using APL is easier than the indirect least-squares approach. Two-stage least squares for the system we used requires two additional APL expressions over ordinary least squares, and that technique yields inconsistent estimates.

An alternative, but related, estimator that we will develop is the instrumental variable estimator. It is related to 2SLS in that the fitted values from the reduced form which we computed can serve as "instruments" for the endogenous variables.

14.3 Instrumental Variables

Instrumental Variables

In those situations in which the conditional distribution of the disturbance term is *not* independent of the regressors for whatever reason (some examples of this problem are "errors in variables" and simultaneous equation models), a useful approach to obtain estimators with desirable statistical properties is that of instrumental variables. More formally, but easier to understand, consider the model

$$\underset{(N \times 1)}{Y} = \underset{(N \times K)}{X} \ \underset{(K \times 1)}{B} + \underset{(N \times 1)}{U}$$

where the dimensions of the arrays are given in parentheses below each array. The main difference between this model and that of Chapter 13 is that here we assume that the distribution of the disturbance term U is *not* independent of X, while in Chapter 13 we did assume such independence.

Now if (and it is a big IF), we can find another matrix Z of dimension $(N \times K)$ such that U is independent of Z *and* the columns of Z have positive coefficients of determination with the columns of X, then Z is said to be a matrix of instrumental variables for X. You might care to remember that the nonindependence of X and U may be caused by only one column of X, say x_1, in which case the variables $x_2, \ldots, x_k$ can "serve as their own instruments."

Well, if we have an instrumental matrix Z, how do we use it? The easy way to do it is this: transform the regression $Y = XB + U$ to

$$Z'Y = Z'XB + Z'U$$

and now carry out an ordinary least-squares regression of $Z'Y$ on $Z'X$ where, with proper choice of Z, $Z'X$, a $(K \times K)$ matrix, is nonsingular. The mathematical solution is

$$\begin{aligned} BE \text{ (the estimator of } B) &= (Z'X)^{-1}Z'Y \\ &= B + (Z'X)^{-1}Z'U \end{aligned}$$

The corresponding covariance matrix of the regression coefficient estimators is defined by

$$\sigma^2(Z'X)^{-1}(Z'Z)(X'Z)^{-1}$$

where σ^2 is the variance of the disturbance term U. The required APL expressions are

```
BE←((⍉Z)+.×Y)⌹(ZX←(⍉Z)+.×X)

VARU←(+/(UH←(Y-X+.×BE))*2)÷-/(⍴X)

COVBE←VARU×(⌹ZX)+.×(⍉Z+.×Z)+.×⌹(⍉ZX)
```

In order to illustrate the use of instrumental variables, we will write a very simple and straightforward routine to do the necessary calculations, but we'll omit all the frills that we put into *MREGRESS;* these you can add later. Consider

```
     ∇BE←INSTRVAR
[1]  BE←((⍉Z)+.×Y)⌹(ZX←(⍉Z)+.×X)
[2]  VARU←(+/(UH←(Y-X+.×BE))*2)÷-/(⍴X)
[3]  COVBE←VARU×(⌹ZX)+.×((⍉Z)+.×Z)+.×⌹(⍉ZX)
[4]  ∇
```

We will check the routine with some data cited by Kmenta, page 313.

```
      Y
0.768 0.433 0.4575 0.5002 0.3462 0.3068 0.3787 ¯0.1188 ¯0.1379
```

```
¯0.2001 ¯0.3845
      X
3.5459 3.2367 3.2865 3.3202 3.1585 3.1529 3.2101 2.6066
2.4872 2.428 2.318
      Z
3.4241 3.1748 3.1686 3.2989 3.1742 3.0492 3.1175 2.5681
2.5682 2.6364 2.5703
      X← ⍉2 11⍴ ((11 ⍴1),X)
      X
1     3.5459
1     3.2367
1     3.2865
1     3.3202
1     3.1585
1     3.1529
1     3.2101
1     2.6066
1     2.4872
1     2.428
1     2.318
      Z← ⍉2 11⍴(( 11 ⍴1),Z)
      Z
1     3.4241
1     3.1748
1     3.1686
1     3.2989
1     3.1742
1     3.0492
1     3.1175
1     2.5681
1     2.5682
```

```
1        2.6364

1        2.5703

         INSTRVAR

¯2.297911012  0.8435302294

         VARU

0.002081788327

         COVBE

0.01070128568     ¯0.00353069422

¯0.00353069422     0.001185860302
```

14.4 Aitken's Generalized Least Squares

The next estimator that we consider is Aitken's generalized least-squares estimator. Suppose that the model we want to estimate is

$$Y = XB + U$$

and the distribution of U has a null mean vector (of dimension N) and a covariance matrix Σ which is of dimension $(N \times N)$ and, in general, has nonzero terms off the diagonal. Suppose Σ is known. If so, an appropriate estimator in such a situation is Aitken's generalized least-squares estimator. It is defined mathematically by

$$B = (X'\Sigma^{-1}X)^{-1}X'\Sigma^{-1}Y$$

with corresponding covariance matrix

$$(X'\Sigma^{-1}X)^{-1}$$

A routine that calculates this estimator is very simple. Some suitable data are

```
         Y

160 160 180 200 210 220 230 250 200 220 230 300 310 340 350

300 400 450 540

         X

2000 2000 2000 2000 2000 2000 2000 2000 4000 4000 4000 4000

4000 4000 4000 6000 6000 6000 6000 6000
```

SIG is a list whose elements are the values of the diagonal elements of Σ

```
         SIG

0.125 0.125 0.125 0.125 0.125 0.125 0.125 0.125 0.1428571429
```

```
0.1428571429 0.1428571429 0.1428571429 0.1428571429
0.1428571429 0.1428571429 0.2 0.2 0.2 0.2 0.2
```

These data are from Kmenta, page 259.

```
      Z←⍉Z←2 20 ⍴Z←Y,X
      Z
160 2000
160 2000
180 2000
200 2000
210 2000
220 2000
230 2000
250 2000
200 4000
220 4000
230 4000
300 4000
310 4000
340 4000
350 4000
300 6000
300 6000
400 6000
450 6000
540 6000
```

A useful routine for calculating Aitken's generalized least squares is

Aitken's Generalized Least Squares

```
      ∇AITKNGLS[⎕]∇
      ∇BE←Z AITKNGLS SIG
[1]   Y←Z[;1]
[2]   X←Z[;(⍳(⍴Z)[2]-1)+1]
[3]   NN←N×N← ⍴ SIG
[4]  (N,N)⍴(NN ⍴1,N ⍴0)\SIG
```

```
[5]    BE←(COVBE←⌹(⍉ X)+.×( ⌹ SIG)+.×X)+.×(⍉X)+.×( ⌹SIG)+.×Y
       ∇
       Z AITNKGLS SIG
0.07315178571
```

The only bit of APL programming which is a little different is in line [4]. What this line does is to produce a diagonal matrix with the elements of the list *SIG* on the main diagonal. `(N N ⍴ 1,N ⍴ 0)`produces a list of length N^2 of 1's and 0's such that when it is used to expand SIG the N^2 elements are the rows of the required diagonal matrix written out in a list.

The Aitken's generalized least-squares estimator is mathematically equivalent to the estimator obtained by ordinary least squares on

$$P'Y = P'XB + P'U$$

where P' is an $N \times N$ matrix which satisfies $P'\Sigma P = I$. Such a P can always be found when Σ is nonsingular.

Another linear regression situation for which a different estimator is required occurs when the regression coefficients are known to satisfy certain linear constraints. The model is

Regression with Restricted Coefficients

$$Y = XB + U$$

$$r = RB$$

where the $(Q \times 1)$ vector r is known, as is the $(Q \times K)$ matrix of coefficients R. In this situation the mathematical expression for the restricted least squares estimator is

$$BER = BE + (X'X)^{-1}R'[R(X'X)^{-1}R']^{-1}(r - (R)BE)$$

where BE is the unrestricted ordinary least-squares estimator and BER is the restricted least-squares estimator. If V represents the covariance matrix of the unrestricted estimator, then the covariance matrix of the restricted estimator is given mathematically by

$$COVBER = V - VR'(RVR')^{-1}RV$$

Expressing these functions in APL is, of course, by now not difficult. For example:

```
       ∇BER←RESLS
[1]    B←Y⌹X
[2]    Q←⌹R+.×(XIX←⌹ (⍉X)+.×X)×.×⍉R
[3]    BER←B+XIX+.× (⍉R)+.×Q+.× (R2-R+.×B)
[4]    VARU←(+/(Y-X+.×BER)*2)÷ (-/(⍴X))
[5]    COVBER←VARU×(XIX-XIX+.×(⍉R +.×(⌹R)+.×XIX+.×⍉R)+.×R+.×XIX)
```

Let's use the following data to try the above function:

```
      Y
100 106 107 120 110 116 123 133 137
      X
1     1        1
1     1.04     0.99
1     1.06     1.1
1     1.11     1.26
1     1.11     1.13
1     1.15     1.03
1     1.2      1.02
1     1.24     1.03
1     1.26     0.98
      R
1 0 0
1 1 0
      R2
1 0
      RESLS
1 ¯1 109.6135479
      VARU
373.1395307
      COVBER
 2.651315829E¯12   ¯3.976973744E¯12    0.000000000E0
¯2.651315829E¯12    4.391241842E¯12   ¯1.988486872E¯12
 0.000000000E0     ¯1.077097056E¯12    3.666426234E1
```

14.5 Durbin's Estimator in First Order Autoregressive Models

All of the estimators so far have been defined in terms of some linear transformation of the regressand vector Y; i.e., *all* the estimators have been of the type

$$B = AY$$

where A is some suitable matrix of *known* constants.

First Order Autoregressive Model

We come now to some estimators which are no longer linear in this sense and which involve iterated solutions. The basic idea is that we start with some reasonable idea for A, say A^0, and from that we get

$$B^1 = A^0Y$$

We use B^1 to calculate an A^1, from which we get

$$B^2 = A^1Y$$

and so on. Usually, it is convenient and statistically justified to stop after obtaining B^2. The iterated nature of such estimators is sometimes not clear from the formulas in the textbooks, especially when one can obtain B^2 as a function (through A^1) of A^0 and Y directly. Our first example of such an iterated estimator is in the context of a time series model.

The model we are considering is

$$Y = XB + U$$

where Y is $(T \times 1)$, X is $(T \times K)$, B is $(K \times 1)$, and U is $(T \times 1)$. The model is similar to our previous models except that the covariance matrix of U is given by:

$$V = \sigma^2 \begin{bmatrix} 1 & \rho & \rho^2 & \cdots & \rho^{T-1} \\ \rho & 1 & & & \rho^{T-2} \\ \cdot & & & & \cdot \\ \cdot & & & & \cdot \\ \cdot & & & & \cdot \\ \rho^{T-1} & \rho^{T-2} & \cdots & \cdots & 1 \end{bmatrix} = \sigma^2\Sigma$$

where ρ is the autocorrelation parameter, $-1 < \rho < 1$; i.e., we are assuming that $U_t = \rho U_{t-1} + e_t$, where e_t is normally distributed with zero mean and constant variance, and e_t, e_s for $t \neq s$ are statistically independent.

There are several methods of estimating a model of this type when ρ is unknown. (Of course, if ρ *is* known, then Aitken's generalized least squares, which was discussed above, can be used.) One of the best methods is a procedure due to Durbin (see, for example, the discussion in Johnston*). The regression model can be transformed to

$$Y_t = Y_{t-1}\rho + X_tB - X_{t-1}B\rho + E_t$$

where Y_t denotes the array Y with the *first* observation deleted, Y_{t-1} denotes the array Y with the *last* observation deleted, X_t is the X matrix with the first row deleted, and X_{t-1} is the X matrix with the last row deleted. E_t is the error term. There are now only $T - 1$ observations. The regression is a two-step procedure.

First Step: Use ordinary least squares to get an estimate r for ρ in $Y_t = Y_{t-1}\rho + X_tB - X_{t-1}B\rho + E_t$

Second Step: (i) Transform variables to get $(Y_t - rY_{t-1})$, $(X_t - rX_{t-1})$; (ii) Use ordinary least squares on the *transformed* vari-

* J. Johnston, *Econometric Methods,* McGraw-Hill, New York, pp. 263–264. 2nd Ed., 1972.

ables to get an estimate for B; i.e., use $(Y_t - rY_{t-1}) = (X_t - rX_{t-1})B + E_t$.

In terms of the APL programming of the routine, we need to obtain the "lagged" arrays Y_{t-1}, X_{t-1}, perform an ordinary least squares (OLS) regressions, transform variables, and do another OLS regression. Consider the following effort:

```
      ∇BE←Y AUTOREG X
[1]     YL←(L1←((((ρ Y)-1)ρ1),0))/Y       {Initial lagging of variables in
                                           lines [1] to [4].
[2]     Y← ( L2←(0,((ρ Y)-1)ρ 1))/Y
[2.2]   K←2+ (ρ X)[2]×2
[3]     XL←L1⌿X
[4]     X←L2⌿X
[5]     BE←Y⌹ (YL, ((ρYL) ρ1),XL,X)       {First step regression, line [5].
[6]     R←BE[1]
[7]     Y←(Y-R×YL)                         {Transform variables, lines
                                           [6] to [8].
[8]     X← ( X-R×XL)
[9]     BE←Y⌹X                             {Second step regression [9].
[.5]  ⍝X MATRIX IS ASSUMED TO BE WITHOUT CONST.
[10]    ∇
```

Let's try an example. In order to see the difference between the estimates, calculate `Y⌹X` with the original variables. First of all, you might wish to check that `(YL,((ρ YL)ρ 1),XL,X)` gives the expected matrix. Let's reuse some familiar data.

```
      Y
1 1.06 1.07 1.2 1.1 1.16 1.23 1.22 1.27
      ρ Y
9
      X
1         1
2.04      0.99
1.06      1.1
1.11      1.26
1.11      1.13
```

```
1.15      1.03
1.2       1.02
1.13      1.03
1.26      0.98
      ρ X
9 2
      YL
1 1.06 1.07 1.2 1.1 1.16 1.23 1.33
      Y
1 1.06 1.07 1.2 1.1 1.16 1.23 1.33 1.37
      XL
1          1
1.04       0.99
1.06       1.1
1.11       1.26
1.11       1.13
1.15       1.03
1.2        1.02
1.24       1.03
      L1
1 1 1 1 1 1 1 1 0
      L2
0 1 1 1 1 1 1 1 1
      (YL, ((ρ YL) ρ1),XL,X)
1        1        1        1        1.04       0.99
1.06     1        1.04     0.99     1.06       1.1
1.07     1        1.06     1.1      1.11       1.26
1.2      1        1.11     1.26     1.11       1.26
1.1      1        1.11     1.13     1.15       1.03
1.16     1        1.15     1.03     1.2        1.02
1.23     1        1.2      1.02     1.24       1.03
```

The result from the routine *AUTOREG* is simple:

```
      Y AUTOREG X
1.16889455    0.1677267818
```

14.6 k-Class Estimators in Simultaneous Equation Systems (OLS, 2 SLS, and Limited Information Maximum Likelihood)

k-Class Estimators

The last estimator to be discussed in this chapter is in fact a class of estimators—the ''k'' class estimators to be precise—which occur in the estimation of regression models embedded within a system of simultaneous equations. The linear regression model is now written in the form

$$\underset{(T \times 1)}{Y} = \underset{(T \times M_1)}{Y_1} \underset{(M_1 \times 1)}{G_1} + \underset{(T \times K_1)}{X_1} \underset{(K_1 \times 1)}{B_1} + \underset{(T \times 1)}{U}$$

where the dimensions of the arrays are put in parentheses below the arrays. Because the variables Y and Y_1 are *jointly* determined by a system of simultaneous equations, U is not statistically independent of Y_1. Y_1 is known as the matrix of included endogenous variables this equation, X_1 is the matrix of included exogenous variables, and X_2 is a $(T \times K_2)$ matrix of the exogenous variables which is called the matrix of excluded (from equation 1) exogenous variables. We will make the simplifying assumption that $K_2 \geq M_1$. Let $X = (X_1X_2)$, a $T \times (K_1 + K_2)$ matrix.

With such a model, an entire class of estimators can be defined by the normal equations as follows:

$$\begin{bmatrix} Y_1'Y_1 - k\hat{V}_1'\hat{V}_1 & Y_1'X_1 \\ X_1'Y_1 & X_1'X_1 \end{bmatrix} \begin{bmatrix} \hat{G}_1(k) \\ \hat{B}_1(k) \end{bmatrix} = \begin{bmatrix} Y_1'Y - k\hat{V}_1'Y \\ X_1'Y \end{bmatrix}$$

where $\hat{V}_1 = Y_1 - \hat{Y}_1$, $\hat{Y}_1 = X(X'X)^{-1}X_1'\ Y_1$, and $\hat{G}_1(k)$, $\hat{B}_1(k)$ are the k-class estimators; for the algebraic details, see Kmenta, Goldberger, or any intermediate level econometrics textbook. If $k = 0$, one has the OLS estimator; if $k = 1$, the two-stage least-squares estimator, and so on. The value of k need not be a preassigned constant, but could be a solution to a maximization problem involving the random variables, as, for example, in the case with the limited information maximum likelihood procedure.

In setting up a program to obtain k-class estimators, let's consider some changes to the approach we have taken so far.

Let's assume that stored in the computer is a matrix Z of dimension $T \times NV$, where NV is the number of variables in the system of equations to be analyzed. Alternatively, and equivalently in APL, the data could be stored as NV arrays of length T. The routine to be defined below will be an interactive routine in that the routine, once it is called, will request input from the user and process it accordingly.

The main steps in the calculation of the k-class estimators are:

(i) Fix the value of k and determine which columns of Z are Y, Y_1, X_1, and X_2;

(ii) Calculate V_1;
(iii) Set up the matrices appearing in the normal equations above;
(iv) Obtain the estimates $\hat{G}_1(k)$, $\hat{B}_1(k)$.

Consider the following routine:

```
     ∇ KCLASSEST;IND;Z;X2;Y1;VH1;Y;Q;P
[1]    'ENTER NO. OF ARRAYS OF VARIABLES'
[2]    NV←⎕
[3]    L1←L2←L3←L4←NV ρ0
[4]    'ENTER ARRAY NO. OF DEPENDENT VB L.'
[5]    ND←⎕
[6]    L1[ND]←1
[7]    'ENTER ARRAY NOS. OF ENDOGENOUS REGRESSOR VB LS.'
[8]    NEND←⎕
[9]    L2[NEND]←1
[10]   'ENTER ARRAY NOS. OF EXOGENOUS INCL REGRESSOR VB LS.'
[11]   NEX←⎕
[12]   L3[NEX]←1
[13]   'ENTER ARRAY NOS. OF EXOGENOUS EXCL. REGRESSOR VB LS.'
[14]   NEXX←⎕
[15]   L4[NEXX]←1
[16]   L4[NEX]←1
[17]   'ENTER DATA ARRAYS ';NV;' IN NO. '
[18]   Z←⎕
[19]   VH1←Y1-X2+.×((Y1←L2/Z)⌹(X2←L4/Z))
[20]   'ENTER VALUE OF K FOR K-CLASS EST.'
[21]   K←⎕
[22]   Y←(((⍉Y1)+.×(L1/Z))-K×(⍉VH1)+.×(L1/Z)),[1]((⍉L3/Z)+.×L1/Z)
[23]   Q←(((⍉Y1)+.×Y1)-K×(⍉VH1)+.×VH1),[2]((⍉Y1)+.×L3/Z)
[24]   P←((⍉L3/Z)+.×Y1),[2]((⍉L3/Z)+.×L3/Z)
[25]   BE←( ,Y)⌹(Q,[1]P)
[26]   BE
     ∇
```

Some discussion is needed for this routine, as it appears to be a little complicated. However, the complications arise from only two sources: the desire to pick out easily which arrays are to be the dependent variables, which the included endogenous, and so on; and the wish to define a whole class of estimators at once. To see how simple two-stage least squares would be without the first requirement (having already removed the second), consider the following pair of lines:

```
YH1←X2+.×(YEND⌹X2)
BE←Y⌹(YH1,X1)
```

where *YEND* is the matrix of included endogenous variables, $X2$ is the matrix of excluded exogenous variables, Y is the dependent variable array, and $X1$ is the matrix of included exogenous regressors. This simple pair of statements generates 2SLS regression coefficient estimates if the appropriate arrays are already specified.

Let's return to the more complicated routine. The first step is to enter the number of variables; since we are going to extract Y, $Y1$, $X1$, and $X2$ from Z by compression (i.e., use of /). For example, $L2$ is an array (of dimension NV) of 1's and 0's—1's where the columns of $Y1$ are located and 0's elsewhere. The device `NV←⎕`, used to solicit a response from the user, has been discussed previously.

In line [18] the user is prompted to insert data in the form of a matrix Z. You can do this by catenating the variable arrays which we have assumed are already stored in the computer. For example, if Q, P, R, S, W are the arrays needed, you type (after the computer prompts with ⎕)

```
⎕ : ⍉NV Tρ(Q,P,R,S,W)
```

where T is the numbers of observations. This action creates a matrix Z whose *columns* are Q, P, etc.

Line [19] calculates $\hat{V}_1$. Lines [22] and [23], [24] define the arrays shown in the normal equations whose solution yields the required k-class regression coefficient estimates; this is done in line [25].

Line [22] is a straightforward programming version of the mathematical statement wherein the two parts of the Y array are catenated together. Lines [23], [24], and [25] are a little more tricky in that we need the extension of catenation called laminate, which we discussed in Chapter 12. The two-dimensional array Q contains $(Y_1' Y_1 - k\hat{V}_1' \hat{V}_1, Y_1'X_1)$, and P contains $(X_1' Y_1, X_1'X_1)$, while `(Q,[1]P)` yields the required matrix:

$$\begin{bmatrix} Y_1' Y_1 - kV_1' V_1 & Y_1'X_1 \\ X_1'Y_1 & X_1'X_1 \end{bmatrix}$$

Line [25] raises another interesting little facet in APL. In the use of ⌹ dyadically, the left argument must be a *list* of length n if the right-hand argument is a two-dimensional array of dimensions $(n \times q)$. The left argument must *not* be a two-dimensional array of dimensions $(n \times 1)$ or $(1 \times n)$. Thus, given the way in which the variable Y was created, $(,Y)$ produces the appropriately dimensioned array.

To test our function, let us consider some data from Kmenta (pages 653–65). The model, which we have used before, is

$$Q_t = a_1 + a_2P_t + a_3D_t + U_{dt} \qquad \text{(Demand Equation)}$$

$$Q_t = b_1 + b_2P_t + b_3F_t + b_4A_t + U_{st} \qquad \text{(Supply Equation)}$$

Here Q_t is the quantity of food consumed per head per year, P_t is the ratio of food prices to general consumer prices, D_t is disposable income, F_t is the ratio of the previous year's prices received by farmers to general prices, and A_t is time in years. Q and P are regarded as endogenous; D, F, and A are exogenous. The demand equation is overidentified, and the supply equation is exactly identified. The data are given in Table 14.1. These data are simulated so that we know the true model, which is

$$Q_t = 96.5 - 0.25\ P_t + 0.30\ D_t + U_{dt}$$

$$Q_t = 62.5 + 0.15\ P_t + 0.20\ F_t + 0.36\ A_t + U_{st}$$

Let us use our k-class function with $k = 0$ (gives OLS) and $k = 1$ (gives 2SLS). In this demand equation, P_t is Y_1 and $i' = (1, 1, \ldots, 1)$, the D_t are X_1, and (F_t, A_t) are X_2. We begin by entering the arrays Q_t, P_t, D_t, F_t, and A_t into the computer.

The output is self-explanatory.

Table 14.1 Data List for Test of k-Class Estimators Routine[a]

Y or Q_t (Dependent Variable)	Y_1 or P_t	X_1 or D_t	X_2 or F_t	X_2 or A_t	X_1 or i
98.485	100.323	87.4	98	1	1
99.187	104.264	97.6	99.1	2	1
102.163	103.435	96.7	99.1	3	1
101.504	104.506	98.2	98.1	4	1
104.24	98.001	99.8	110.8	5	1
103.243	99.456	100.5	108.2	6	1
103.993	101.066	103.2	105.6	7	1
99.9	104.763	107.8	109.8	8	1
100.35	96.446	96.6	108.7	9	1
102.82	91.228	88.9	100.6	10	1
95.435	93.085	75.1	81	11	1
92.424	98.801	76.9	68.6	12	1
94.535	102.908	84.6	70.9	13	1
98.757	98.756	90.6	81.4	14	1
105.797	95.119	103.1	102.3	15	1
100.225	98.451	105.1	105	16	1
103.522	86.498	96.4	110.5	17	1
99.929	104.016	104.4	92.5	18	1
105.223	105.769	110.7	89.3	19	1
106.232	113.49	127.1	93	20	1

[a] These data are stored for us (the authors) in an array called $K565$.

First round—Ordinary Least-Squares Estimators ($K = 0$):

```
      KCLASSEST
ENTER NO. OF ARRAYS OF VARIABLES
⎕:
      6
ENTER ARRAY NO. OF DEPENDENT VBL.
⎕:
      1
ENTER ARRAY NOS. OF ENDOGENOUS REGRESSOR VBLS.
⎕:
      2
ENTER ARRAY NOS. OF EXOGENOUS INCL REGRESSOR VBLS.
⎕:
      3 6
ENTER ARRAY NOS. OF EXOGENOUS EXCL. REGRESSOR VBLS.
⎕:
      4 5
ENTER DATA ARRAYS 6 IN NO.
⎕:
      K565
ENTER VALUE OF K FOR K-CLASS EST.
⎕:
      0
¯0.3162988049   0.3346355982   99.89542291
```

Second Round—2SLS Estimators ($K = 1$):

```
      KCLASSEST
ENTER NO. OF ARRAYS OF VARIABLES
⎕:
      6
ENTER ARRAY NO. OF DEPENDENT VBL.
```

```
⎕:
      1
ENTER ARRAY NOS. OF ENDOGENOUS REGRESSOR VBLS.
⎕:
      2
ENTER ARRAY NOS. OF EXOGENOUS INCL REGRESSOR VBLS.
⎕:
      3 6
ENTER ARRAY NOS. OF EXOGENOUS EXCL. REGRESSOR VBLS.
⎕:
      4 5
ENTER DATA ARRAYS 6 IN NO. ⎕
⎕:
      K565
ENTER VALUE OF K FOR K-CLASS EST.
⎕:
      1
¯0.2435565378    0.3139917943   94.63330387
```

Third Round—Limited Information Maximum Likelihood Estimators (K = 1.1739):

```
      KCLASSEST
ENTER NO. OF ARRAYS OF VARIABLES
⎕:
      6
ENTER ARRAY NO. OF DEPENDENT VBL.
⎕:
      1
ENTER ARRAY NOS. OF ENDOGENOUS REGRESSOR VBLS.
⎕:
      2
```

```
ENTER ARRAY NOS. OF EXOGENOUS INCL REGRESSOR VBLS.
⎕:
      3 6
ENTER ARRAY NOS. OF EXOGENOUS EXCL. REGRESSOR VBLS.
⎕:
      4 5
ENTER DATA ARRAYS 6 IN NO.
⎕:
      K565
ENTER VALUE OF K FOR K-CLASS EST.
⎕:
      .739
¯0.2295353985    0.3100126821    93.61902556
```

Limited Information Maximum Likelihood Estimators

The value of k in the *L.I.M.L.* estimator is obtained by minimizing the ratio of the error sum of squares of the regression of $(Y - Y_1G_1)$ on X_1 and on X; for an explanation, see Kmenta (page 569).

We have now reached the end of the book. We trust that you have discovered by this stage the power (and beauty) of APL and, more importantly, that in using APL you learned statistics more easily and fully than would otherwise have been possible.

There is more to learn, in APL statistics and econometrics. To do that you will need to read some more books. As far as APL is concerned, you might now find it useful and interesting to look through some of the computer manuals recommended in the bibliography. The statistics and econometrics books listed are all excellent at their various and respective levels of sophistication and detail.

Summary

This chapter discussed a number of APL routines for calculating regression estimators in some interesting extensions to the simple model. The first topic is that of simultaneous equations. The program to estimate the general k-class estimators was given. In addition, the 2SLS, indirect least squares, and instrumental variables estimators were discussed at length.

The next topic to be discussed was Aitken's generalized least-squares procedure, which is needed to estimate regression models where the covariance matrix of the disturbance terms is nonscalar.

Another useful extension was to consider Durbin's method for estimating a linear regression model, where the disturbance term was distributed according to a first order autoregressive process.

Exercises

Statistical Applications

1. Write in APL the following formulas useful in simultaneous equation systems.

   (a) $Z = \Sigma_{n\times n} \otimes I_{K\times K}$, where $\otimes$ is the mathematical Kronecker product.

   (b) $B = [Z'(\hat{\Sigma}^{-1}\otimes X(X'X)^{-1}X')Z]^{-1}[\hat{\Sigma}^{-1}\otimes X(X'X)^{-1}X']Y$

   (c) $B = [Z[I\otimes X(X'X)^{-1}X']Z]^{-1}Z[I\otimes X(X'X)^{-1}]Y$

2. Let the true model be $Y = 2X_1 + 3X_2 + U_t$, where

$$X = [X_1, X_2] = \begin{bmatrix} 0.2 & 0.1 \\ 0.3 & 0.8 \\ 0.4 & 0.6 \\ 0.5 & 0.5 \\ 0.6 & 0.5 \\ 0.7 & 0.4 \end{bmatrix}$$

   The U_t are identically and independently distributed random variables with mean zero and variance 1 (see the solution for a routine that generates independent random numbers from a normal distribution with zero mean and variance equal to $N(0, 1)$).

   (a) Given X and U, find the values of Y.

   (b) Compute the variance of Y using the computed values of Y.

3. Consider the following data:

Y	X_1	X_2
49	35	55
40	38	60
46	40	70
45	41	61
52	42	68
59	38	71
55	31	59
61	44	66
64	38	75
65	29	85

   where Y = the production, in thousands of bushels, of wheat of one farm over a period of 10 years; X_1 = mean January temperature; and X_2 = mean June temperature.
   Let the relationship between the variables be linear and of the form $Y = a_0 + a_1X_1 + a_2X_2 + U$.

Test the following hypotheses:

$$H_0: \quad a_2 = 0$$
$$\text{against } H_1: \quad a_2 \neq 0$$
and
$$H_0: \quad a_0 = a_1 = a_2 = 0$$
$$\text{against } H_1: \quad a_0, a_1, a_2 \neq 0$$

Use a 5% significance level for your tests.

4. Referring to exercise 3, suppose you believe that the June temperature affects the production twice as much as does the January temperature, and you want to test if your belief is true. Mathematically, this can be solved by the following steps (given without proof).

   (a) Find $\hat{a} = (X'X)^{-1}X'Y$ — the unconstrained vector of a.

   (b) Find $\bar{a} = \hat{a} + W(r - R\hat{a})$ — the restricted vector of a.
   where $R = [0 \quad 2 \quad -1]$, $r = 0$
   $W = (X'X)^{-1}R'A$
   $A = [R(X'X)^{-1}R']^{-1}$ — Notice $Ra = r \Leftrightarrow a_2 = 2a_1$

   (c) Find $\hat{\lambda} = AR(\hat{a} - \bar{a})$ and $V\hat{\lambda} = \sigma_u^2 A$, the variance-covariance matrix of λ, where σ_u^2 is the variance of the residuals from the restricted regression.* Thus under H_0: $a_2 = 2a_1$ $\lambda \sim N(0, V_\lambda)$, which is equivalent to testing

   $$H_0: \quad \lambda = 0$$
   $$\text{against } H_1: \quad \lambda \neq 0$$

   (d) Another statistic that you might want to use is

   $$\frac{\bar{u}'\bar{u} - \hat{u}'\hat{u}}{\hat{u}'\hat{u}} \cdot \frac{n - k - 1}{s} \sim F_{s,n-k-1}$$

   where $\bar{u}'\bar{u}$ is the restricted error sum of squares, $\hat{u}'\hat{u}$ is the unrestricted error sum of squares, and s is the number of constraints.

   (e) Yet another statistic that you might want to use is

   $$-2 \ln \frac{\hat{u}'\hat{u}}{\bar{u}'\bar{u}}$$

   which is distributed as chi-square with 5 degrees of freedom.

   (f) Do you get the same answers using the three different tests? That is, do you accept or reject your hypotheses in all three tests?

5. Consider the following simple market model:

   $$q^s = 2 + p + u \qquad q^d = 4 - p + v$$

   where q^s is quantity supplied, q^d is quantity demanded, the equilibrium condition is $q^s = q^d$, and $u \sim N(0, 3)$ and $v \sim N(0, 1)$, where u and v are independent.

* λ is the Lagrange multiplier of the minimization problem $L = u'u + 2\lambda(Ra-r)$.

(a) Find the expected equilibrium price.

(b) Suppose that the government initiates an inquiry whenever the equilibrium price is more than \$1.50. Find the probability of having the government make an inquiry in this market.

6. Refer to exercise 3. Suppose you believe that there is no constant term in that relationship, and you run the regression

$$Y = a_1X_1 + a_2X_2 + U$$

(a) Find the error sum of squares of this regression and compare it to error sum of squares of the regression in exercise 3.

(b) Suppose that instead of running the previous regression, you decide to run

$$Y = a_1V + a_2X_2 + U^*$$

where V is the residual vector of the regression of X_1 on X_2, including a constant term. Use the same data to show that the estimate of a_2 is the same as that given by the regression of Y on X_2.

7. Write a program that will calculate the following statistics in the heteroskedastic model, which is given by $Y = XB + U$, where the covariance matrix of U is Σ, which is diagonal with unequal elements on the diagonal.

(a) $B = (X'\Sigma^{-1}X)^{-1}X'\Sigma^{-1}Y$, the efficient estimator of the vector of regression coefficients, where Σ is a diagonal matrix whose elements are known in advance.

(b) The t-statistics for the B estimates.

(c) Use the regression routine you have just written together with the data set *WATT* (Appendix E) to estimate the regression

$$Y = a_0 + a_1X_1 + a_2P_1 + U$$

where Y is the KWHR of the 15 plants for the first year, X is the BTU of the 15 plants for the first year, and P_1 is the vector of prices that the 15 plants paid during the first year. Σ is a diagonal matrix with the numbers 1 to 15 on its diagonal; the numbers correspond to the size of the plant (1 for the smallest, 15 for the largest) measured by the total KWHR produced during the first year.

8. Consider the following model that determines the equilibrium price and quantity of a product.

$$D_t = a_0 - a_1P_t + a_2Y_t + U_{1t}$$

$$S_t = b_0 + b_1P_t + U_{2t}$$

where D_t = quantity demanded, endogenous; P_t = price of product, endogenous; S_t = quantity supplied, endogenous; Y_{2t} = disposable income, exogenous; $D_t = S_t$, the equilibrium condition.
U_{1t}, U_{2t} are each identically and independently distributed normal

variables with means zero, variances σ_1^2 and σ_2^2, respectively, and covariance σ_{12}.

(a) Solve for the reduced form.

(b) Is OLS an appropriate method of estimation for the reduced form coefficients?

(c) Generate by computer a sample of observations on U_{1t}, U_{2t} and Y_t for 20 observations and compute values for D_t, S_t, and P_t by solving the equations after picking values for the coefficients. Use your generated data to run OLS and 2SLS regressions on both equations and comment on your results.

9. Consider the very simple income determination model:

$$C_t = a_0 + a_1 Y_t + U_t$$

$$Y_t = C_t + I_t$$

where C_t = aggregate consumption, I_t = Investment (exogenous), Y_t = GNP, U_t = stochastic disturbance.
Use the data set MACRO in Appendix E.

(a) Estimate a_0 and a_1 using OLS on equation (1).

(b) Write the reduced form equation for C, (using equations 1 & 2).

(c) Estimate a_0 and a_1 using OLS on the reduced form equation for C.

10. Consider the following product market model for the United States. Use the data set *MACRO*. See Appendix E. Can you estimate these equations?

$$C_t = a_1 Y_t + a_2 RL_{t-1} + U_{1t}$$
$$I_t = b_1 Y_t + b_2 RL_{t-1} + U_{2t}$$
$$Y_t = C_t + I_t$$

Check to see which, if any, of these equations are identified.

11. Using the money supply series in data set MACRO, estimate the parameters of the model:

$$ML_t = a_0 + a_1 ML_{t-1} + U_t$$

12. A partial adjustment hypothesis could state that the target level of per capita consumption (C_t^*) is a linear function of income, i.e.:

$$C_t^* = a + bY_t$$

Actual consumption is adjusted towards C^* according to

$$C_t - C_{t-1} = \gamma[C_t^* - C_{t-1}] + U_t, 0 < \gamma < 1$$

which means that people adjust their consumption towards the desired level. However, the adjustment is not generally accomplished in one period ($\gamma < 1$). Use the data in MACRO from 1950 to 1977 in Appendix E to test this hypothesis. On the basis of your estimated model predict consumption for the year 1978 and check your prediction with the actual 1978 value of consumption.

13. Suppose that you want to test Milton Friedman's permanent income hypothesis. You first want to translate his hypothesis into mathematical formulas and test it on real data. The entire model might be summarized in the following two equations:

$$C_t = a + bY\rho_t$$

$$Y\rho_t = \gamma\Sigma_{i=0}^{\infty}(1 - \gamma)^i Y_{t-i}$$

These equations can be combined to yield the following final equation:

$$C_t = a\gamma + b\gamma Y_t + (1 - \gamma)C_{t-1} + U_t$$

Verify that this final equation is identical to the model in exercise 12 after substitution of C_t^* and reinterpret the estimated coefficients.

Appendix A

The Computer: Where It Is and How to Get Access to It

A.1 Account Number and Password

In this brief appendix you will learn how to ''log-on'' to the computer. First, let's repeat that every computer center has its own administrative procedures for determining who can use the machine and what resources are allocated to those users. Some organizations make these arrangements for you, otherwise you go to the computer center to do it yourself. In any event, it's nothing to be overly concerned about. Typically, a form is completed and you are given an ''account number'' and a ''password,'' and sometimes a budget or time constraint. The account numbers are usually six or fewer digits, and the passwords are eight or fewer characters (either digits or letters). As an example, at Stanford the ''account numbers'' have two parts: a user number-J66 for the project, and a user group-E1. The password was HØV. (To discourage poachers, a slang term for people who specialize in using other people's accounts for their own purposes—such as playing space war—one should keep his or her password confidential.)

Account No. and Password

With this administrative chore done, you are ready to begin. First find an unoccupied computer terminal. The keyboard looks like any electric typewriter, and in fact the letters are in the same position. The end of this appendix contains diagrams of some typical keyboards—one for an IBM 5120 minicomputer, an IBM-2741 terminal and a DEC-Writer II, and a Hewlett-Packard-HP2541A—literally hundreds of others exist, although most have the standard keyboard layout and features. In other respects, each terminal is slightly different; they hide the on-off switch in different places, provide extra buttons, extra keys, lights, and even electronic whistles. None of these matters should worry you, but you must be sure that the terminal has the special APL characters. If your terminal does not have these special symbols, it is probably called an ASCII (American Society for Coded Information Interchange) terminal. These terminals do not support the special character set of APL; however, all is not lost. The APL language allows you to use mnemonic codes in place of the special symbols. Look at the letter R; above it is the Greek letter rho, ρ. To represent ρ you would type the mnemonic code .RO and the computer would understand it to be the symbol ρ. We have included a list of mnemonic codes used in APL in Appendix C. These vary somewhat from computer to computer, and you should check with your facility if you must use one of these ASCII terminals. You are now ready to start, and we presume that you are sitting in front of the terminal.

A.2 Log-on Procedure

On the IBM-5120 minicomputer there is a white switch on the front control panel marked APL/BASIC; push it to the APL position and turn the red power (on-off) switch on. That is all there is to it. When the machine warms up, the TV screen will say *CLEAR WS*, meaning that you are ready to go with a clear workspace—the electronic equivalent of a blank piece of paper. If you have a 5120 or similar microcomputer, you can skip to the log-off section of this appendix.

Those of you who are operating in a time-sharing environment will have to use a more complicated log-on procedure. On the 5120, which is a small desktop minicomputer, there is one user. On a big computer, hundreds of users are connected to the computer. These connections are done in one of two ways: via a direct wire from your terminal to the computer, or over the telephone system. If you see a telephone by every terminal, you know how your terminal is connected.

Unfortunately the log-on procedure varies from computer center to computer center. However, the procedures are usually not difficult to acquire. We suggest that if you are not provided with a log-on procedure sheet by the computer center, you make one up yourself until you have everything memorized. Believe us, one of the most frustrating parts of programming is learning how to get programming access to the computer; once that is done, the rest is plain sailing. The problem here is that, as we mentioned above, every center has its own ideas about how people should log-on.

Log-on Procedures

Nevertheless, to give you an idea as to what is involved, consider the following procedure. Starting with the direct wire situation, you turn on the terminal and type the symbol), then your account number and password, and then press the "return" key. This button sends one line of information to the computer. This could look like

```
)1984:ME
```

your account being number 1984 and your password ME. This password is sometimes called a "LOCK." The computer will respond with something like:

```
062*    10:41:32    9/10/80    Von Mises

OPR:  SYSTEM AVAILABLE TO 22:30
```

CLEAR WS

Translating this into English, it means that you are using telephone line number 62 (called a ''port'' by computernics), the time is 10:41 and 32 seconds A.M. on the tenth day of September, 1980, and the user's name is Von Mises. The operator indicates that the system will be available until 10:30 tonight.

If, when you turned on your terminal, you could not type on it, or when you typed nothing happened, it might be that the terminal is in local mode. Look for a switch marked COM/LOC for communication and local. It needs to be in the COM mode to communicate with the computer. Clear Work Space means you are ready to begin.

If you connect to the computer via a telephone line, you have exactly the same log-on procedure, with the additional task of calling the computer. The procedure is simple enough; first you activate a small box called a modem. Modem stands for Modulator Demodulator. This modem essentially takes the signals from your terminal and sends them over the telephone line, and then retranslates the signals from the computer to your terminal.

These modems are of two basic types: one a Data-Phone and the other an acoustic coupler. The Data-Phone is usually a Western Electric device; you simply pick up the telephone receiver, press the ''talk'' button, and dial the computer's telephone number after you get a dial tone. When the computer electronically answers the phone, you will hear a high pitched tone that indicates that the computer is ready for you. You then press the ''data'' button.

The computer will typically send a brief message or just one character—a − or a *. You can now ''log-on'' exactly as the direct-wired terminal user does.

The final connecting device is the acoustic coupler. Again the terminal power is turned on. With this equipment, the electric power to the modem is also turned on, and the telephone number is called. When you hear the high pitched tone, push the telephone hand set into the coupler receptacle. Sometimes there is an indication of which direction to place the hand set, but if there isn't, just take a guess. If it doesn't work, try it the other way around. Again, you will get some message from the computer and you log-on.

As an example, here is how it would be done at the Stanford Computer Facility. Our connection is hard wired, so we turn on the terminal (a GenCom, with the switch cleverly placed under the keyboard so that it is almost impossible to find). I type an ''a'' and the computer responds with a +, I type SCF APL and the machine responds . . . SCF 168 . . . waits a moment and then types User?. I type J66; it responds, Group? I tell it El; it asks, Password? I reveal *HØV* (since changed) and it asks, Command?. I respond CALL APL; the machine types APL, then skips a line and types *CLEAR WS*.

As another example, consider the sign-on procedure used at New York University when using the CUNY computer center. At this point we run into another option on terminals—some can be used for both APL and other languages. The character set for other languages is often designated STD CHAR SET (Standard Character Set), and you should be in standard to begin. The procedure is:

Switch on power of terminal and modem;

Ensure character switch is set to STD;

Dial up and place phone in acoustic coupler when high pitched tone comes on line (a white light on the acoustic coupler comes on if the connection is good);

Type "shift" P

Hit RETURN key (sometimes called EXECUTE);

System responds: type: a for apl, w for wylbur, o for callos

Type: a

Hit RETURN key;

System responds: PROJECT No., ID?

Type: [the project number given to you by the computer center];

System responds: PASSWORD?

Type: [the password you have picked and which is on file with the computer center];

Change character switch from STD to ALT;

Type:)BLot [the ")" symbol is upper case] key on the APL keyboard (see end of this appendix)]

System responds: XXXXXXXXXXXXXXXXXXXXXXX

Type:) followed by APL account number, the symbol : and password;

System responds:

Date Your Name [as filed in computer center]
CUNY APL

Your log-on is now complete, and you are ready to begin.

By now you realize that every center is different and will have its own special way of logging you onto the machine. However, you now know enough to understand the general idea of the log-on procedures used in any computer center. They all have handouts that explain the details and peculiarities of their own system.

A.3 Log-off Procedure

Logging off is simple. On the microcomputer systems you just turn the machine off. Of course, everything in the machine will be erased if you do not copy it onto a tape or disk. The same thing is true when you use the terminals. You must make special provision to save anything you want to use again. You learned how to do this in Chapter 7. If you have a 5120 or other microcomputer, you can skip to the last sentence of this appendix.

To log-off on the terminal, you type)ØFF and press the return button. The machine will respond, indicating that it understood your command. The response is something like:

```
062*      13:01:04   9/10/80    Von Mises
Started   10:41:32  Clocktime this session 2:22:32  to date 17:25:30
                    CPUTime this session 0:00:10  to date 00:01:10
```

Log-off Procedure

The first line here is similar to the first line of your terminal session; it indicates your port, the current time (1:04 P.M.), the date, and your name. The second line repeats the time you started and computes the elapsed

IBM 5120 desktop computer showing the APL character set, numeric pad, and special function keys.

This top photograph showing a 5120 desktop computer can be programmed in either Basic or APL with the flip of a switch. The keyboard is exactly like a standard typewriter in that pressing the shift key (either of the keys with the wide arrows on the bottom rank of keys) results in the APL characters being entered into the computer. A convenient feature is that by holding the command key (CMD, on the far left) and pressing one of the keys on the top row will produce an entire command. For example, holding down CMD and pressing 1 results in the command) LOAD being entered automatically.

IBM 5120 showing keyboard characters that can be entered using the command key

The lower photograph shows the special overstruck characters that can be produced with one stroke. The command is held down and any of the individual keys now represents a new symbol or combination of key strokes. For example, pressing the CMD key and the F key results in the divide quad or domino function being entered. If the machine were in the Basic programming mode the characters input would have been entered. Using the CMD key saves a number of key strokes and is a handy feature.

Operator uses Hewlett-Packard 2641A CRT terminal specifically designed for APL and featuring extensive communications capabilities.

The operator is using the H-P 2641A terminal as a device to communicate with a central computer. In the upper right corner above the keyboard are two tape cassettes for data.

Closeup of Hewlett-Packard 2641A terminal keyboard showing the various APL characters and special function keys.

The closeup photo of the Hewlett-Packard 2641A CRT terminal keyboard shows the APL characters, numeric pad, cursor control, and tape control keys. The dial in the upper left corner allows the operator to select the communications rate from 110 to 9600 band. The higher rates are used for direct or hardwired communications to the host computer and the lower rates are generally used when communication is over telephone lines.

time that you were connected to the machine for this session and all your previous sessions. The third line tells how much computer time you have used.

The final steps are to disconnect the terminal from the computer and turn off the electricity to the equipment. If your terminal is hard wired or connected via a DataPhone, just turn the electricity off. If the terminal is connected via an acoustic coupler, you turn off the electricity to the terminal, hang up the phone, and turn off the power to the coupler.

These log-off procedures also vary from installation to installation. At Stanford, to log-off you would type: *)OFF*, and press the return button. (Remember—the computer does not get your message until you send it by pressing the return button.) The computer responds: Command? Next, the user types "logoff," and the computer responds with a large amount of information including the CPU time, elapsed time, the dollar charges to your account, and the words END OF SESSION.

IBM 5120 showing the full keyboard and dual diskette slots.

Gerald Musgrave

The IBM has the full APL character set and two diskette drives. The diskettes are used to store APL functions and data.

Pictured is the MCM System 900 desktop APL computer, manufactured by MCM Computers Ltd. of Kingston, Ontario, Canada and distributed in the United States by Interactive Computer Systems, Inc., New York, New York.

The MCM 900 is a desktop computer that can be programmed in APL as a standalone device. It can also be used as a communications terminal to other computers using its data communication interface. A number of applications routines are available to assist the APL programmer in the areas of business management, pension administration, bilingual word processing, and financial accounting.

Appendix B

Longley Benchmark

The "Longley Benchmark," as it has come to be called, was developed in a paper titled "An Appraisal of the Least Squares Programs for the Electronic Computer from the Point of View of the User."[1] The results were shocking to some since a number of well-known and extensively used programs produced inaccurate results. In fact some of the computed coefficients have the wrong sign! Beaton and Barone used the same data in

OLS Regression Equation

$$Y = B_0 + B_1X_1 + B_2X_2 + B_3X_3 + B_4X_4 + B_5X_5 + B_6X_6 + u$$

Regressand	Regressor Variables					
Y Total Derived Employment	X_1 GNP Implicit Price Deflator 1954 = 100	X_2 Gross National Product	X_3 Unemployment	X_4 Size of Armed Forces	X_5 Noninstitutional Population 14 Years of Age and Over	X_6 Time
60,323	83.0	234,289	2,356	1,590	107,608	1947
61,122	88.5	259,426	2,325	1,456	108,632	1948
60,171	88.2	258,054	3,682	1,616	109,773	1949
61,187	89.5	284,599	3,351	1,650	110,929	1950
63,221	96.2	328,975	2,099	3,099	112,075	1951
63,639	98.1	346,999	1,932	3,594	113,270	1952
64,989	99.0	365,385	1,870	3,547	115,094	1953
63,761	100.0	363,112	3,578	3,350	116,219	1954
66,019	101.2	397,469	2,904	3,048	117,388	1955
67,857	104.6	419,180	2,822	2,857	118,734	1956
68,169	108.4	442,769	2,936	2,798	120,445	1957
66,513	110.8	444,546	4,681	2,637	121,950	1958
68,655	112.6	482,704	3,813	2,552	123,366	1959
69,564	114.2	502,601	3,931	2,514	125,368	1960
69,331	115.7	518,173	4,806	2,572	127,852	1961
70,551	116.9	554,894	4,007	2,827	130,081	1962

[1] Longley, James W., *American Statistical Association Journal,* Vol. 62, No. 31, 7 September, 1967, pp. 819–841.

other programs, and they found similar inaccuracies.[2] As a special check, they wrote their own routine, using multiple precision arithmetic and were careful about the numerical analysis methods they used. Their results were in agreement with Longley's calculations. So we have some confidence in the computed accuracy of the benchmark. We should add parenthetically that Beaton and Barone caution researchers on interpreting regression results in problems of this nature. However, our point is that, *ceterus paribus,* more accuracy should be preferred to less.

We have included the raw data and some of the published results, and the results from our APL runs with the ⌹ function.

Computed Coefficients

	Longley Computation Using Desk Calculator	Longley Computation Using High Precision Program 360	APL-STAT
B_0	−3482258.6330	−3482258.634597	−3,482,258.634594932
B_1	15.061872271	15.06187227161	15.06187227114229
B_2	−0.035819179	−0.035819179293	−0.03581917929260994
B_3	−2.020229803	−2.020229803818	−2.020229802817922
B_4	−1.033226867	−1.033226867174	−1.033226867173836
B_5	−0.051104105	−0.051104105653	−0.05110410565236916
B_6	1829.15146461	1829.151461614112	1829.151464613044

Standard Error of Regression Coefficient

	Longley	APL-STAT
X_0	890420.3836	890420.3862191836
X_1	84.9149	84.91492578619182
X_2	0.0335	0.03349100783754301
X_3	0.4884	0.4883996826502304
X_4	0.2143	0.2142741633587326
X_5	0.2261	0.2260732001756643
X_6	455.4785	455.4785004777639

Computed *t*-Values

	Longley	APL-STAT
t_0	−3.9108	−3.91080290668244
t_1	0.1774	0.1773760282034131
t_2	−1.0695	−1.0695163151363
t_3	−4.1364	−4.136427347486052
t_4	−4.8220	−4.821985306012059
t_5	−0.2261	−0.2260511445525611
t_6	4.0159	4.015889800933297

This should give you an idea of our results compared to the benchmark. You might want to try this problem on your computer with the *APL-STAT* routines and also with other program packages.

[2] Beaton, A., Rubin, D., and Barone, J., "The Acceptability of Regression Solutions: Another Look at Computational Accuracy," *Journal of the American Statistical Association,* Vol. 71, No. 353, March, 1976, pp. 158–168.

Appendix C

APL Character Set

Single-Strike Characters

APL Set	ASCII Set	Mnemonic	Name
+	+		add
A-*Z*	A-Z		alphabetics
∧	&	.AN	ANd
←	← or _		assignment
,	,		concatenate, comma
:	:		colon
.	.		decimal point
÷	%	.DV	DiVide
=	=		equal to
\	\		expand
⋆	*		exponentiate
>	>	.GT	Greater Than
[	[		left bracket
(	(		left parenthesis
<	<	.LT	Less Than
×	#		multiply
0-9	0-9		numerics
'	'		quote string
?	?		question (roll and deal)
/	/		reduce
]	]		right bracket
)	)		right parenthesis
;	;		semicolon
-	-		subtract
↑	∧	.TK	TaKe
_		.US	UnderScore
|		.AB	residue (ABsolute value)
α		.AL	ALpha
⎕		.BX	quad (BoX)
⌈		.CE	CEiling (maximum)

Overstruck Characters

APL Set	ASCII Set	Mnemonics	Name
⌽		.RV	ReVersal
⍉		.TR	TRanspose
⍎		.XQ	eXecute
⍕		.FM	ForMat
A̲-Z̲		.ZA-.ZZ	underscored alphabetics
⍙		.Z@	underscored lower del

Single-Strike Characters

APL Set	ASCII Set	Mnemonic	Name
↓		.DA	drop (Down Arrow)
¨		.DD	Dieresis
⊥		.DE	DEcode
∇	$	.DL	DeL
⋄		.DM	DiaMond
∩		.DU	Down Union
⊤		.EN	ENcode
∊		.EP	EPsilon
⌊		.FL	FLoor
≥		.GE	Greater than or Equal
→		.GO	GO to (branch)
⍳		.IO	IOta
{		.LB	Left Brace
∆		.LD	delta (Lower Del)
≤		.LE	Less than or Equal
⊢		.LK	Left tacK
○		.LO	circle (Large O)
⊃		.LU	Left Union
≠		.NE	Not Equal to
¯		.NG	NeGation
∼		.NT	NoT
⍵		.OM	OMega
∨		.OR	OR
}		.RB	Right Brace
⊣		.RK	Right tacK
⍴		.RO	RhO
⊂		.RU	Right Union
∘		.SO	jot (Small o)
∪		.UU	Up Union

Overstruck Characters

APL Set	ASCII Set	Mnemonic	Name
⍝	"		LAmp (Comment)
!	!		factorial
$	$	.DS	dollar sign
⍒		.GD	Grade Down
⍋		.GU	Grade Up
⌶		.IB	I-Beam
⍟		.LG	LoGarithm
⍲		.NN	NaNd
⍱		.NR	NoR
⍀		.CB	Column expansion
⊖		.CR	Column Rotate
⌿		.CS	Column Reduction
⌹		.DQ	Divide Quad
O		.OU	OUt
⍫		.PD	Protected Del

Overstruck Characters

NAME	CHARACTER	KEYS
Comment	⍝	∩ C, ∘ J
Compress*	⌿	\ /, - +
Execute	⍎	⊥ B, ∘ J
Expand*	⍀	\ /, - +
Factorial, Combination	!	' K, : .
Format	⍕	⊤ N, ∘ J
Grade Down	⍒	∇ G, | M
Grade Up	⍋	∆ H, | M
Logarithm	⍟	* P, ○ O
Matrix Division	⌹	⎕ L, ÷ X
Nand	⍲	∧ 0, ~ T
Nor	⍱	∨ 9, ~ T
Protected Function	⍫	∇ G, ~ T
Quad Quote	⍞	⎕ L, ' K
Rotate, Reverse	⌽	| M, ○ O
Rotate, Reverse*	⊖	○ O, - +
Transpose	⍉	\ /, ○ O

* These are variations of the symbols for these functions. These variations are used when the function is to act on the first coordinate of an array.

Appendix D

Saving Your Workspace on The IBM 5110 Microcomputer

If you want to save your workspace and you are using the 5120, you will use the cartridge tape. You must first "mark" the number of tape files on the tape and then transfer the workspace onto the tape. The details of marking the tape are not directly relevant for this discussion. However, if you type

```
)MARK 32 2 1
```

you will mark 2 tape files. The numbering of the files begins with one, and each file contains 32 blocks of 1024 Bytes. There is enough room to store the whole memory of a machine with 32,768 Bytes of core. The computer will respond with

```
MARKED 0002 0032
```

To copy the workspace onto the tape, key in

```
)CONTINUE 1001  PROB
```

The first 1 indicates tape drive 1 (on the machine), 001 means you are storing the information of the first file and you are naming it *PROB*. The 5110 will respond

```
CONTINUED  1001  PROB
```

if all goes well. You could then take your tape out, turn the machine off and walk away with the tape. Next time, after turning the machine on and obtaining the `CLEAR WS` signal, you would push the tape cartridge into the machine and key

```
)LOAD  1001  PROB
```

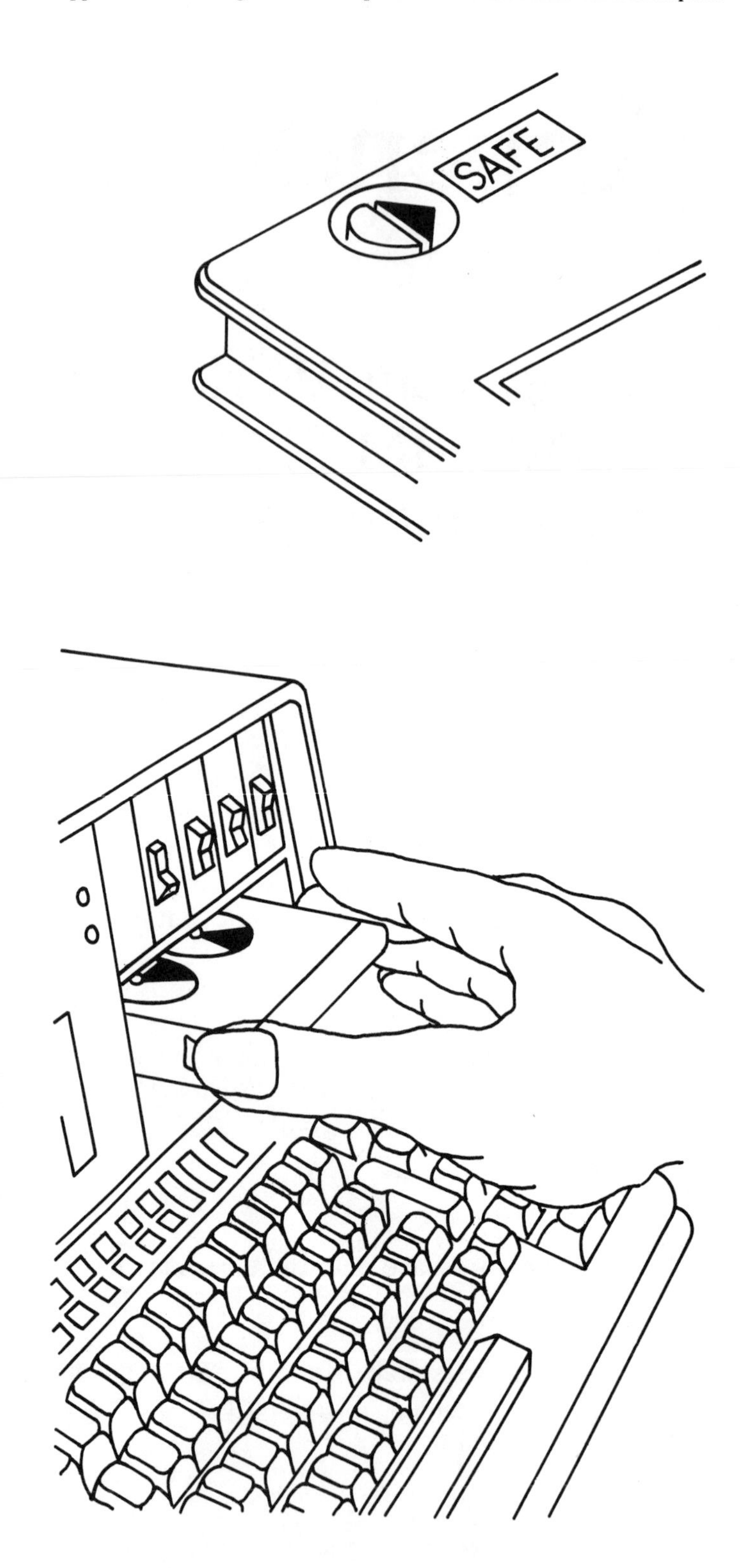
SAFE

The computer responds

    LOADED 1001 PROB

and you are ready to go.

The)SAVE ,)COPY , and)PCOPY commands are used in much the same way as they were described in the previous section. If you skipped that section, you should read it now. The main differences concern marking the tape as you just learned, keeping track of the physical record number of the workspace, and the fact that continued workspaces are not automatically loaded when the computer is turned on.

If you forget the physical location (file number) of your saved workspaces, you can issue the)LIB command. The computer will respond with the file number and the name of the workspace. It also displays other information about the file—but that does not concern us now. Two warnings about this command are appropriate. Firstly, the computer does not automatically rewind the tape to its beginning. You should issue the)REWIND command to accomplish this. Secondly, a small slot, marked with an arrow, exists on the top of the cartridge (see the figure). You must turn the little wheel (a dime in the slot is perfect) until the arrow points away from the word ''safe,'' as in the figure. If the arrow points to ''safe'' the computer will not write on the tape, although you can read from the tape at all times.

On other mini- and microcomputers you may save your functions and variables on floppy disks. This is also true for the IBM 5110 and 5120 series machines. The procedure for disks is similar to the one we outlined for tapes.

Appendix E

Data Set 'Macro'

	Units of Measurement	*Abbreviations*
A. GNP and its components		
1. GNP	Billions of 1972 US $	Y
2. Consumption	Billions of 1972 US $	C
a) Nondurables	Billions of 1972 US $	CN
b) Durables	Billions of 1972 US $	CD
3. Investment	Billions of 1972 US $	I
4. Government	Billions of 1972 US $	G
5. Imports	Billions of 1972 US $	IM
6. Exports	Billions of 1972 US $	X
B. Income		
1. Disposable Income	Billions of 1972 US $	YD
C. Prices, Wages, Interest Rates		
1. GNP Deflator	=100 in 1972	P
2. Short-term interest rates	% per annum	RS
3. Long-term interest rates	% per annum	RL
4. Wage rate	in current US $'s per week	W
D. Employment		
1. Employment	Thousands of persons	L
2. Unemployment	%	U
E. Assets		
1. Capital Stock	Billions of US $	K
F. Money Supply	Billions of US $	ML

Source: Economic report of the President, January 1979. Series only from 1950 to 1978.

Data Set Macro

Year	Y	C	I	G	X	IM	YD
1950	533.5	338.1	93.7	97.7	21.7	17.7	205.5
1951	576.5	342.3	94.1	132.7	25.9	18.5	224.8
1952	598.5	350.9	83.2	159.5	24.9	20.0	236.4
1953	621.8	364.2	85.6	170.0	23.8	21.8	250.7
1954	613.7	370.9	83.4	154.9	25.3	20.8	255.7
1955	654.8	395.1	104.1	150.9	27.9	23.2	273.4
1956	668.8	406.3	102.9	152.4	32.3	25.0	291.3
1957	680.9	414.7	97.2	160.1	34.8	26.0	306.9
1958	679.5	419.0	87.7	169.3	30.7	27.2	317.1
1959	720.4	441.5	107.4	170.7	31.5	30.6	336.1

Data Set Macro (Continued)

Year	Y	C	I	G	X	IM	YD
1960	736.8	453.0	105.4	172.9	35.8	30.3	349.4
1961	755.3	462.2	103.6	182.8	37.0	30.3	362.9
1962	799.1	482.9	117.4	193.1	39.6	33.9	383.9
1963	830.7	501.4	124.5	197.6	42.2	35.0	402.8
1964	874.4	528.7	132.1	202.7	47.8	36.9	437.0
1965	925.9	558.1	150.1	209.6	49.1	41.0	472.2
1966	981.0	586.1	161.3	229.3	51.6	47.3	510.4
1967	1,007.7	603.2	152.7	248.3	54.2	50.7	544.5
1968	1,051.8	633.4	159.5	259.2	58.5	58.9	588.1
1969	1,078.8	655.4	168.0	256.7	62.2	63.5	630.4
1970	1,075.3	668.9	154.7	250.2	67.1	65.7	685.9
1971	1,107.5	691.9	166.8	249.4	67.9	68.5	742.8
1972	1,171.1	733.0	188.3	253.1	72.7	75.9	801.3
1973	1,235.0	767.7	207.2	252.5	87.4	79.9	901.7
1974	1,217.8	760.7	183.6	257.7	93.0	77.1	984.6
1975	1,202.3	774.6	142.6	262.6	90.0	67.5	1,086.7
1976	1,271.0	819.4	173.4	262.8	95.9	80.5	1,184.4
1977	1,332.7	857.7	196.3	269.2	98.2	88.7	1,303.0
1978	1,385.1	891.2	210.1	275.2	107.3	98.7	1,451.2

Data Set Macro (Continued)

Year	P	RL	RS	W	L
1950	53.64	2.62	1.45	53.13	58,918
1951	57.27	2.86	2.16	57.86	59,961
1952	58.00	2.96	2.33	60.65	60,250
1953	58.88	3.20	2.52	63.76	61,179
1954	59.69	2.90	1.58	64.52	60,109
1955	60.98	3.06	2.18	67.72	62,170
1956	62.90	3.36	3.31	70.74	63,799
1957	65.02	3.89	3.81	73.33	64,071
1958	66.06	3.79	2.46	75.08	63,036
1959	67.52	4.38	3.97	78.78	64,630
1960	68.67	4.41	3.85	80.67	65,778
1961	69.28	4.35	2.97	82.60	65,746
1962	70.55	4.33	3.26	85.91	66,702
1963	71.59	4.26	3.55	88.46	67,762
1964	72.71	4.40	3.97	91.33	69,305
1965	74.32	4.49	4.38	95.45	71,088
1966	76.76	5.13	5.55	98.82	72,895
1967	79.02	5.51	5.10	101.84	74,372
1968	82.57	6.18	5.90	107.73	75,920
1969	86.72	7.03	7.83	114.61	77,902
1970	91.36	8.04	7.72	119.83	78,627
1971	96.02	7.39	5.11	127.31	79,120
1972	100.00	7.21	4.69	136.90	81,702
1973	105.80	7.44	8.15	145.39	84,409
1974	116.02	8.57	9.87	154.76	85,935
1975	127.15	8.83	6.33	163.53	84,783
1976	133.76	8.43	5.35	175.45	87,485
1977	141.61	8.02	5.60	188.64	90,546
1978	152.09	8.73	7.99	203.34	94,373

Data Set Macro (Continued)

Year	U	ML	K
1950	5.3	0	0
1951	3.3	0	0
1952	3.0	0	0
1953	2.9	128.8	0
1954	5.5	132.3	0
1955	4.4	135.2	0
1956	4.1	136.9	0
1957	4.3	135.9	0
1958	6.8	141.1	0
1959	5.5	143.4	0
1960	5.5	144.2	0
1961	6.7	148.7	254.7
1962	5.5	150.9	269.7
1963	5.7	156.5	288.2
1964	5.2	163.7	305.6
1965	4.5	171.4	336.0
1966	3.8	175.8	364.0
1967	3.8	187.4	386.2
1968	3.6	202.5	426.5
1969	3.5	209.0	473.6
1970	4.9	219.7	492.3
1971	5.9	234.0	529.6
1972	5.6	255.3	599.3
1973	4.9	270.5	697.8
1974	5.6	282.9	734.6
1975	8.5	295.2	756.3
1976	7.7	313.5	823.1
1977	7.0	338.5	900.1
1978	6.0	361.1	953.6

[0] Not available

Data Set Watt*
List of the Plants

Number	Plant Name
1	J. H. Cambell
2	B. C. Cobb
3	Dan E. Kern
4	J. R. Whiting
5	Rock River (Wis.)
6	Nelson Dewey (Wis.)
7	Minn. Power and Light Co., Clay Boswell
8	St. Clair
9	Presque Isle
10	Gulf Power (Scholtz)
11	Gulf Power (Lansing Smith)
12	Tampa Electric Company (F. J. Gannon)
13	Northern States Power Co. (Lawrence)
14	Montana Dakota Utilities Co. (Heshett)
15	Appalachian Power Co., Clinch River

Source: U.S. Federal Power Commission, Steam Electric Plant Construction Cost and Annual Production Expenses 1966–1977.
U.S. Energy Information Administration 1974.

Data Set Watt

The data set WATT consists of three tables, the KWHR produced in Table 1, the BTU's used in Table 2, and the prices of BTU's in Table 3. The observations are for 15 plants over a period of 10 years. Thus, WATT is a 3 by 10 by 16 (including the year column) three-dimensional array.

Table 1 Number of KWHR in millions

Plant Number

Year	1	2	3	4	5
1966	1,938.4	2,754.9	4,066.6	2,575.7	927.7
1967	2,980.4	2,752.4	3,878.5	2,425.5	977.8
1968	3,904.3	2,923.6	4,055.3	2,442.2	971.7
1969	3,678	2,907.9	3,900.7	2,393.9	987.9
1970	4,273.9	2,901.9	3,590.3	2,276.2	926.5
1971	3,316.7	3,244.4	3,558.2	2,190.1	807.5
1972	3,971.3	2,874.3	3,489.9	1,921.2	827.8
1973	4,103.7	3,449.8	3,438.2	2,211.8	862.4
1974	3,921.4	3,506.8	3,328.9	2,440.5	727.5
1975	3,460.9	2,869.6	3,032.5	2,155.4	546.9

Year	6	7	8	9	10
1966	1,393	892.2	9,792.8	854.2	246.0
1967	1,453.1	1,046.5	9,468.6	1,114.6	233.6
1968	1,505.9	1,023.2	9,373.4	1,135.4	313.4
1969	1,454.9	941.2	11,671.6	1,229.6	415.4
1970	1,275.0	947.9	12,620.4	1,161.4	389.4
1971	1,286.1	916.5	12,606.2	1,086.8	430.9
1972	1,439.8	945.6	12,386.8	1,121.0	433.3
1973	1,371.4	2,007.4	10,509.6	1,136.4	505.9
1974	1,126.5	2,773.6	9,958.2	1,130.8	462.2
1975	1,186.9	3,192.1	8,266.0	1,795.7	489

Year	11	12	13	14	15
1966	1,006.8	4,481.7	128	355.3	5,660.3
1967	1,648.4	4,666.5	176.2	462.5	5,345.6
1968	2,231.9	5,078.1	144.8	510.7	5,191.0
1969	2,173.2	4,954.9	113.2	497.7	5,353.9
1970	1,961.5	5,350.0	124.5	564.8	5,223.7
1971	1,754.6	4,736.7	133.3	583.2	5,052.4
1972	1,827.5	5,136.3	167.7	612.9	5,484.9
1973	2,043.8	4,880.3	168.1	575.9	5,575.2
1974	1,722.8	3,821.6	150.3	610.8	4,974.5
1975	1,592.3	4,392.5	40.8	602.9	4,202.2

Table 2 Number of BTU's in 10^{12} Units

Plant Number

Year	1	2	3	4	5
1966	663.2	1,049.4	1,376.2	986.4	402.9
1967	1,093.6	1,066.4	1,331.9	942.3	413.9
1968	1,429.9	1,133.6	1,426.6	955	425.7
1969	1,349.5	1,138.9	1,384.6	955.2	440.5
1970	1,569.1	1,200.4	1,308.0	922	431.3
1971	1,237.8	1,363.0	1,302.6	913.1	378.1
1972	1,497.0	1,190.3	1,289.7	790.7	389.1
1973	1,540.4	1,424.3	1,282.9	831.3	414.2
1974	1,488.7	1,474.0	1,258.1	972	362.6
1975	1,355.4	1,238.1	1,128.8	838.4	291.4

Table 2 (Continued)

Year	Plant Number 6	7	8	9	10
1966	601	397.8	3,765.6	362.5	118.6
1967	625.6	426.9	3,633.5	454.2	110.2
1968	648.9	411.6	3,592.5	465.9	144.5
1969	631.6	485.5	4,483.7	507.6	192.5
1970	560.4	583.0	4,797.9	487.7	186
1971	565.4	570.2	4,788.3	476.8	218.4
1972	638.1	605.8	4,508.5	511.5	224.3
1973	605.2	1,143.4	3,560.2	516.4	241.6
1974	507.2	1,707.6	473.9	549.4	225
1975	600.8	2,063.0	2,434.0	794.3	235.5

Year	Plant Number 11	12	13	14	15
1966	422.1	1,965.5	18.3	330.2	2,069.5
1967	699.6	2,113.3	28.1	425.1	1,964.8
1968	941.5	2,340.2	16	467.3	1,912.5
1969	930.1	2,285.8	26.5	463.2	1,990.1
1970	868	2,466.3	31.8	520.2	1,972.0
1971	770.9	2,213.4	46.9	541.9	1,917.9
1972	849.0	2,420.6	52.6	571.3	2,068.5
1973	893.3	2,246.3	47.7	542.3	2,145.4
1974	771.3	1,808.1	59.6	575.9	1,948.6
1975	716.2	1,039.9	32.5	577.3	1,609.2

Table 3 Cost per Million BTU in ¢

Year	Plant Number 1	2	3	4	5
1966	30.40	29.40	30.7	27.20	31.79
1967	27.30	29.90	31.80	27.70	32.38
1968	27.60	31.70	31.70	28.40	33.36
1969	29.00	34.00	31.80	30.10	34.94
1970	30.41	39.90	40.20	33.20	40.39
1971	34.10	45.50	46.60	41.60	47.80
1972	37.92	46.75	46.99	43.91	51.99
1973	44.00	49.20	48.34	61.43	55.49
1974	57.70	57.70	66.69	116.67	85.66
1975	111.90	101.30	104.9	142.40	133.90

Year	Plant Number 6	7	8	9	10
1966	26.78	35.54	28.30	32.44	30.01
1967	26.32	35.00	28.50	33.54	29.88
1968	26.62	36.00	29.20	34.29	30.15
1969	28.41	36.00	29.40	35.96	31.96
1970	30.78	32.90	32.90	41.53	36.52
1971	34.11	31.70	39.70	51.30	43.70
1972	38.06	31.85	43.60	56.94	51.50
1973	43.37	31.68	47.71	64.10	54.80
1974	65.49	34.38	86.90	96.98	102.39
1975	96.60	62.00	106.00	134.03	136.83

Table 3 (Continued)

	Plant Number				
Year	11	12	13	14	15
1966	25.07	26.66	38.93	20.95	17.31
1967	25.29	26.13	38.18	20.52	17.71
1968	25.67	26.36	39.23	20.00	18.69
1969	26.99	29.28	41.45	20.42	20.36
1970	30.08	30.25	46.02	21.84	22.11
1971	40.02	31.02	54.09	22.90	26.47
1972	43.97	39.54	56.88	24.60	29.64
1973	46.74	49.80	65.51	25.36	34.50
1974	72.74	68.88	69.93	28.18	72.92
1975	124.40	123.00	135.95	45.50	128.90

Function Glossary

Table 1 Monadic Functions

Symbol	Name	Definition	Example
+	Conjugate	+A is A	A 7 +A 7
-	Negative	-A is 0-A	A 7 -A ¯7
×	Signum	×A is (A>0)-A<0 When A is positive a 1 results. When A is negative a ¯1 results. When A is zero a 0 results.	A 7 ×A 1 ×(-A) ¯1
÷	Reciprocal	÷A is 1÷A	A 7 ÷A 0.14286

Table 1 Monadic Functions (Continued)

Symbol	Name	Definition	Example
|	Magnitude	Absolute Value	B 7.653 ¯7.653 7.456 ¯7.456 |B 7.653 7.653 7.456 7.456
⌊	Floor	Least Integer	B 7.653 ¯7.653 7.456 ¯7.456 ⌊B 7 ¯8 7 ¯8
⌈	Ceiling	Greatest Integer	B 7.653 ¯7.653 7.456 ¯7.456 ⌈B 8 ¯7 8 ¯7
?	Roll	?A is random number from set of (⍴A) consecutive integers with each integer having the (1÷⍴A) probability of being selected.	?A 1 ?7777777 6 4 4 2 1 5 5
*	Exponent	e^A	A 7 *A 1096.6
⍟	Natural Logarithm	ln A or $\log_e A$	A 7 ⍟A 1.9459
○	Pi Times	$\pi \times A$	○1 3.1416 C 1 2 3 ○C 3.1416 6.2832 9.4248

Table 1 Monadic Functions (Continued)

<table>
<tr><th>Symbol</th><th>Name</th><th>Definition</th><th>Example</th></tr>
<tr><td>!</td><td>Factorial</td><td>!A=1×2×...×(A-1)×A</td><td>A
7
!A
5040
!4
24</td></tr>
<tr><td>~</td><td>Not</td><td>~1 is 0 ,~0 is 1.
Truth table defined for 0 and 1 only.</td><td>D
1 0 1
~D
0 1 0
~~D
1 0 1</td></tr>
</table>

Table 2 Dyadic Scalar Functions

<table>
<tr><th>Symbol</th><th>Name</th><th>Definition</th><th>Example</th></tr>
<tr><td>+</td><td>Plus</td><td>Add</td><td>3 + 1.02
4.02
5 + 4
9</td></tr>
<tr><td>-</td><td>Minus</td><td>Subtract</td><td>6 - 7
¯1</td></tr>
<tr><td>×</td><td>Times</td><td>Multiply</td><td>5 × 4
20</td></tr>
<tr><td>÷</td><td>Divide</td><td>Divide</td><td>5 ÷ 4
1.25</td></tr>
<tr><td>|</td><td>Residue</td><td>Remainder after divide</td><td>5|25.010
0.01</td></tr>
<tr><td>⌊</td><td>Minimum</td><td>Smaller of two values</td><td>5 ⌊ 4
4
¯7.001 ⌊ ¯7.01</td></tr>
</table>

Table 2 Dyadic Scalar Functions (Continued)

Symbol	Name	Definition	Example
			¯7.01
⌈	Maximum	Greater of two values	5 ⌈ 4
			5
			¯7.001 ⌈ ¯7.01
			¯7.001
*	Power	A to the B power: A^B	3 * 2
			9
			4 * .5
			2
⍟	General Logarithm	The base A logarithm of B	10⍟5
			0.69897
			10⍟1005
			3.0022

Symbol	Name	Definition
○	Circular, Hyperbolic, and Pythagorean Functions	1○X = Sine X
		2○X = Cosine X
		3○X = Tangent X
		4○X = (1+X*2)*.5
		5○X = Sinh X
		6○X = Cosh X
		7○X = Tanh X
		¯1○X = Arcsin X
		¯2○X = Arccos X
		¯3○X = Arctan X
		¯4○X = (¯1+X*2)*.5
		¯5○X = Arsinh X
		¯6○X = Arcosh X
		¯7○X = Artanh X
		0○X = (1+X*2)*.5

Table 2 Dyadic Scalar Functions (Continued)

Symbol	Name	Definition	Example
!	Binomial Coefficients	$\binom{A}{B} = \frac{A!}{(A-B)!B!}$ $\binom{10}{5} = 5!10 = 252$	The Bth term in the expansion of $(X + Y)^A$, also the number of combinations of A things taken B at a time.
∧	And		
∨	Or		
⍲	Nand (not and)		
⍱	Nor (not or)		

A	B	A∧B	A∨B	A⍲B	A⍱B
0	0	0	0	1	1
0	1	0	1	1	0
1	0	0	1	1	0
1	1	1	1	0	0

A and B must be logical (0 or 1) variables

Symbol	Name	Definition
<	Less than	Result is 1 if relation holds (TRUE) and 0 if it does not hold (FALSE). For example, 5 < 7 is 1, 5 > 7 is 0.
≤	Not greater than	
=	Equal to	
≥	Not less than	
>	Greater than	
≠	Not equal to	

Table 3 Dyadic Array Functions

Function	Symbol	Scalar f Array	Array f Array
Plus	+	3 + 4 8 ¯2 7 11 1	3 4 1 + 2 4 6 5 8 7
Minus	-	3 - 4 8 ¯2 ¯1 ¯5 5	3 4 1 - 2 4 6 1 0 ¯5
Times	×	3 × 4 8 ¯2 12 24 ¯6	3 4 1 × 2 4 6 6 16 6
Divide	÷	3 ÷ 4 8 ¯2 0.75 0.375 ¯1.5	3 4 1 ÷ 2 4 6 1.5 1 0.16667
Residue			3|4 8 ¯2 1 2 1	3 4 1|2 4 6 2 0 0
Minimum	⌊	3⌊4 8 ¯2 3 3 ¯2	3 4 1⌊2 4 6 2 4 1
Maximum	⌈	3⌈4 8 ¯2 4 8 3	3 4 1⌈2 4 6 3 4 6

Table 3 Dyadic Array Functions (Continued)

Function	Symbol	Scalar *f* Array	Array *f* Array
Power	*	3 * 4 8 ¯2	3 4 1 * 2 4 6
		81 6561 0.1111	9 256 1
Logarithm	⍟	3 ⍟ 4 8 2	3 4 2 ⍟ 2 4 6
		1.2619 1.8928 0.63093	0.63093 1 2.585
Circle	○	3○ 4 8 ¯2	3 4 1 ○ 2 4 6
		1.1578 ¯6.7997 2.185	¯2.185 4.1231 ¯0.27942
Binomial	!	3 ! 4 8 2	3 4 1 ! 2 4 6
		4 56 0	0 1 6
And	∧	1∧1010	1010∧1100
		1010	1000
Or	∨	1∨1010	1010∨1100
		1111	1110
Nand	⍲	1⍲1010	1010⍲1100
		0101	0111
Nor	⍱	1⍱1010	1010⍱1100
		0000	0001
Less	<	3<4 8 2 3	3 4 1<2 4 6
		1 1 0 0	0 0 1
Not greater	≤	3≤4 8 2 3	3 4 1≤2 4 6
		1 1 0 1	0 1 1
Equal	=	3=4 8 2 3	3 4 1 = 2 4 6
		0 0 0 1	0 1 0
Not equal	≠	3≠4 8 2 3	3 4 1 ≠ 2 4 6
		1 1 1 0	1 0 1
Not less	≥	3 ≥ 4 8 2 3	3 4 1 ≥ 2 4 6
		0 0 1 1	1 1 0
Greater	>	3 > 4 8 2 3	3 4 1 > 2 4 6
		0 0 1 0	1 0 0

Table 4 Mixed Functions

Name	Form	Definition	Examples Arrays used in Examples:
Shape or Size	ρA	Results in vector whose elements are the number of elements in A if A is a vector, or the dimension of A if A is an array.	A B C 1 2 3 10 20 100 200 4 5 6 30 40 300 7 8 9 50 60 70 80 ρC 3 ρA 3 3 ρB 4 2
Ravel	,A	Results in vector whose elements are the elements of A in row order.	,A 1 2 3 4 5 6 7 8 9
Reshape	AρB	Reshapes the ravel of B to shape specified by A.	2 2 ρB 10 20 30 40 2 4 ρA 1 2 3 4 5 6 7 8 3 5 ρA 1 2 3 4 5 6 7 8 9 1 2 3 4 5 6
Reversal	⌽A or ⊖A	Reverses elements in A. ⌽ reverses the elements along the last coordinate. ⊖ reverses the elements along the first coordinate.	⌽A 3 2 1 6 5 4 9 8 7 ⊖A 7 8 9 4 5 6

Table 4 Mixed Functions (Continued)

Name	Form	Definition	Examples Arrays used in Examples:
			1 2 3 ⌽C 300 200 100
Rotate	A⌽B or A⊖B	The elements of *B* rotated *A* positions. ⌽ rotates elements along the last coordinate and ⊖ rotates elements along the first coordinate.	2⌽C 300 200 100 2⌽A 3 1 2 6 4 5 9 7 8 2⊖B 50 60 70 80 10 20 30 40
Transpose	⍉A or A⍉B	⍉A transposes the axes of array *A*. A⍉B, arranges axes of *B* to conform to argument *A*.	⍉A 1 4 7 2 5 8 3 6 9 ⍉B 10 30 50 70 20 40 60 80 1 1⌽A 1 5 9 2 1 ⍉A 1 4 7 2 5 8 3 6 9 1 2 ⍉A 1 2 3 4 5 6

Table 4 Mixed Functions (Continued)

Name	Form	Definition	Examples Arrays used in Examples:
			7 8 9
Catenate	A,A A,[I]B	Joins two arrays along last axis. A,[I]B joins B to A on the Ith axis.	C,ι5 100 200 300 1 2 3 4 5 B,B B,[10 20 10 20 10 20 30 40 30 40 30 40 50 60 50 60 50 60 70 80 70 80 70 80 10 20 30 40 50 60 70 80
Laminate	A,[J]B	Joins two arrays along a new axis, where J is not an integer, new axis is ⌈J.	C,[.1]C 100 200 300 100 200 300 C,[1.1]C 100 100 200 200 300 300 ρ(A,[1.1]A) 3 2 3 A,[1.1]A 1 2 3 1 2 3 4 5 6 4 5 6 7 8 9 7 8 9

Table 5 Mixed Functions

Name	Form	Definition	Example
Take	N↑A	If *N* is positive, *first N* elements are taken and if negative, last *N* elements are taken from vector *A*.	2↑C 100 200 ¯1↑C 300 2 2↑A 1 2 4 5
Drop	N↓A	If *N* is positive, first *N* elements are dropped and if negative, last *N* elements are dropped from vector *A*.	2↓C 300 ¯1↓C 100 200 2 2↓A 9
Compress	N/A	Selects elements from *A* as determined by zero one argument *N*. For each 1 in *N*, the corresponding element in *A* is selected and for each 0, the element is not selected.	0 1 0/C 200 1 0 0/A 1 4 7 1 0 0/[1]A 1 2 3
Expand	N\A	Fills array with alphabetic spaces or numeric zeros corresponding to zeros in the argument *N*.	1 0 1 1\C 100 0 200 300 1 0 1 0 1\[1]A 1 2 3 0 0 0 4 5 6 0 0 0 7 8 9

Table 5 Mixed Functions (Continued)

Name	Form	Definition	Example
Indexing	A[]	Selects elements from A depending on expression enclosed in brackets.	C[1 3] 100 300 A[1;3] 3 A[1 2 3;3] 3 6 9 B[;2] 20 40 60 80
Index of	A ⍳ B	Returns the index value of first occurrence of B in A.	1 2 3 4 5 6 7⍳7 7 0 0 1 1⍳1 3 2 3 4 2 1⍳1 5 3 5 6 2
Index Generator	⍳ A	Generates first A integers in order.	⍳10 1 2 3 4 5 6 7 8 9 10
Membership	A∊B	Determines if each element of A is a member of B.	2 3 4 2 1∊2 1 0 0 1 0 2 3 4 2 1∊2 3 1 1 0 1 0
Grade Up	⍋A	Returns the index values of A in ascending order.	⍋4 2 3 1 5 4 2 3 1 5
	A[⍋A]	Sorts the elements of vector A in ascending order.	⍋ 40 20 31 10 55 4 2 3 1 5 S[⍋S←4 2 3 1 5] 1 2 3 4 5
Grade Down	⍒A	Returns the index of values of A in descending order.	⍒4 2 3 1 5 5 1 3 2 4 ⍒40 20 30 30 10 55

Table 5 Mixed Functions (Continued)

Name	Form	Definition	Example
	A[⍒*A*]	Sorts the elements of vector *A* in descending order.	6 1 3 4 2 5 *S*[⍒*S*←40 20 30 30 10 55] 55 40 30 30 20 10
Deal	*A*?*B*	Selects *A* random integers without replacement from ⍳*B*, each integer has a (1÷⍴*B*) chance of selection.	2?20 18 5 10?10 6 5 7 8 2 10 1 9 4 3 10?10 10 9 4 3 6 8 5 2 1 7
Matrix Inverse	⌹*A*	Produces the inverse of a nonsingular matrix.	*RND*←2 2 ⍴4⍳4 *RND* 1 3 4 2 ⌹*RND* ¯2 1.5 1 ¯0.5
Domino	*B*⌹*A*	Domino can be used to solve a set of linear equations if *A* has the same number of rows as columns. Domino returns the coefficient of least-squares regression if the number of rows of *A* exceeds the number of columns.	$5X_1 + X_2 + 3X_3 = 7$ $10X_1 + 3X_2 + 5X_3 = 10$ $^{-}2X_1 + 0.2X_2 + 8.3X_3 = 0$ *A* 5 1 3 10 3 5 ¯2 0.2 8.3 *B*←7 1 0 0 *B*⌹*A* 1.7881 ¯3.4851 0.51485

Table 5 Mixed Functions (Continued)

Name	Form	Definition	Example
			AA
			5 1 3
			10 3 5
Domino ⌹			¯2 .2 8.3
			2 ¯3.5 0.5
			B
			7 1 0 0 1
			B⌹AA forces intercept through zero
			0.90461 0.20213 0.22486
			B⌹1,AA catenate column of ones for intercept
			6.2918 0.26813 1.5607 ¯0.7310
Quad output	⎕←A	Displays A and generates a line feed.	⎕←'CHARACTER'
			CHARACTER
			⎕←SUMS←+/B
			18
Quote quad output	⍞←A	Displays A with no line feed.	In a Function
			⍞←'A1&&&'.
			⍞←'CHARACTER'
			A1&&&CHARACTER
Quad Input	A←⎕	Enters a line of input from a function.	A←⎕
			⎕:
			1 2 3
			A
			1 2 3
Quote quad input	A←⍞	Reads a line of characters and creates a character vector from inside a function.	A←⍞
			_
			A C B
			A
			A C B

Table 5 Mixed Functions (Continued)

Name	Form	Definition	Example
Format monadic	⍕A	Monadic format A results in a character representation of A so it can be catenated & displayed on same line with character data.	('THE LOG OF',⍕A),'EQUALS',⍕⍟A THE LOG OF 2 EQUALS 0.69315
Format dyadic	A⍕B	Displays B according to specification A. The first element of A is the number of columns and the second element is the print precision.	A 5 1 3 10 3 5 ¯2 0.2 8.3 2 ¯3.5 0.5 6 3⍕A 5.000 1.000 3.000 10.000 3.000 5.000 ¯2.000 .200 8.300 2.000 ¯3.500 .500 7 1 7 2 7 3⍕A 5.0 1.00 3.000 10.0 3.00 5.000 ¯2.0 .20 8.300 2.0 ¯3.50 .500

Bibliography

Barron, D. W., *Recursive Techniques in Programming,* American Elsevier, New York, 1968

Beaton, A., Rubin, D., and Barone, J., *'The Acceptability of Regression Solutions: Another Look at Computational Accuracy,' Journal of the American Statistical Association,* Vol. 71, No. 353, March 1976, pp. 158–168.

Dhrymes, Phoebus J., *Distributed Lags: Problems of Estimation and Formulation,* Holden-Day Inc., San Francisco, 1971.

Dhrymes, Phoebus J., *Introductory Econometrics,* Springer-Verlag: Berlin-Heidleberg-New York, 1979.

Gilman, L., and Rose, A., *APL An Interactive Approach,* John Wiley and Sons, New York, 1976.

Hamburg, Morris, *Statistical Analysis for Decision Making,* Harcourt, Brace and World, Inc., New York, 1970, pp. 347–348.

IBM-APL Reference Manuals; APL/360 User's Manual (GH20-0906), APL Primer (GH20-0689), APL Language (GC26-3847) IBM Corp., New York, 1977.

IBM 5110 APL Reference Manual, SA21-9303-1, IBM Corp., (General System Division), Atlanta, 1978.

IBM Systems APL Language, GC26-3847-3, IBM Corp. (Programming Publishing), San Jose, 1978.

Iverson, K. E., *A Programming Language,* John Wiley and Sons, New York, 1962.

Johnston, J., *Econometric Methods,* McGraw Hill, New York, 1972.

Kaplan, W., *Advanced Calculus,* Addison-Wesley, Boston, 1952.

Kmenta, J., *Elements of Econometrics,* MacMillan, New York, 1977.

Longley, James W., 'An Appraisal of the Least Squares Programs for the Electronic Computer from the Point of View of the User,' *American Statistical Association Journal,* Vol. 62, No. 317, pp. 819–841, September 1967.

Mendenhall, W., and Reinmuth, J., *Statistics for Management and Economics,* Wadsworth, Belmont, California, 1978.

Neter, J., Wasserman, W., Whitmore, G., *Applied Statistics,* Allyn and Bacon, Boston, 1978.

Pearson, E. S., and H. O. Hartley, *Biometrika Tables for Statisticians,* Vol. 1, Cambridge University Press, 1962, p. 104.

Press, James, *Applied Multivariate Analysis,* Holt, Rinehart and Winston: New York, 1972.

Rao, R. C., *Linear Statistical Inference and Its Applications,* John Wiley and Sons, New York, 1968.

Smillie, K. W., *Stat Pac I,* Department of Computing Service, University of Alberta, Edmonton, Alberta, Canada, 1968.

Smillie, K. W., *Statpack2: an APL Statistical Package,* Department of Computing Science, University of Alberta, Edmonton, Alberta, Canada, June 1969.

Theil, H., *Introduction to Econometrics,* Prentice-Hall, New Jersey, 1978.

Wonnacott, T., and Wonnacott, R., *Introductory Statistics for Business and Economics,* John Wiley and Sons, New York, 1977.

Answers to Exercises

In this section a few answers to selected problems are presented. A complete set of solutions to both the APL exercises and statistical problems is available from the publisher.

2

Statistical Applications

1. 15.00
2. 13.33
3. 91.125
4. 0.2457E⁻18
5. (a) 254.97166
   (b) 14.99
   (c) 5.02
   (d) 0.057656
6. 36,383.88

3

APL Practice

1. (a) blank
   (c) 0
   (e) 3
   (h) 3
2. (a) 1 1 1 1
   (d) 1 4 9 16
   (f) Domain error
   (k) 0 1 1.58 2
   (p) Domain error
   (r) 2.5
   (u) 100
   (v) 1 ⁻1 2 ⁻2
3. F1
   ⁻44 ⁻18 0 6 10 12 12
   F3
   552 56 0 2 0 6 56
4. (a) 2
   (b) 7 15
   (e) 1.83
5. (a) 2.928
   (d) 3085
   (g) ⁻5
   (j) 110
7. (a) 45
   (e) 5.5
   (h) 6.2E20
   (l) 82.999 82.5

Statistical Applications

1. a. 380.4
   c. 1.0905
   e. 0.08515

2. a. 4.2

4. a. 7.3%
b. 0.8

6. a. $.8
b. 2.88 and 5.99

7. $194.32

4. c. 0.447
d. (1) 19.61, 20.51
(2) 19.16, 20.95
(3) 18.72, 21.40

5. d. 2
e. 0.6
f. zero
i. $S_3 = {}^{-}0.6$
$S_4 = 2.2$

4

APL Practice

1. (a) 0 0 0 0 0 0 0 0 0 0 0
(e) 6 7 8 9 0 1 2 3 4 5 6
(i) ¯3 ¯2
(l) 0 0 ¯2 ¯1
(o) Index Error

2. (a) 3 random numbers from 1 to 6 with replacement.

4. ((7ρ0),4ρ1)/Z

5. We use the first 20 terms of each series.
(a) 0.69
(e) does not converge
(f) 0.58198
(i) ≈¯.3
(k) does not converge

7. (a) Cauchy-Schwartz
+/X × Y ≤ (+/X*2) × (+/Y*2)

8. (e) type →
(f) you will have an infinite loop.

Statistical Applications

2. The maximum value of S is 5 when $P = .5$ and the minimum value of S is 0 when $P = 1$.

3. Mean equals 5.920
Variance equals 1.920
Standard deviation equals 1.360

5

APL Practice

1. (o) (X[⍋X])[3]

2. (a)
```
¯2 ¯1 ¯1 0 0 1 1 2 2
 2  2  1 1 0 1 1 2 2
 0  1  0 1 0 1 0 1 0
```
(c) Same as Z

4. 3.246 3.246 ¯549

6. (a) $X + 2X + 3X + 4X$
(c) none
(d) I

8. The root is between 0.3169 and 0.3170

Statistical Applications

4. a. Mean = 2.6
Variance = 3.3714
Standard Deviation = 1.8361
Mean Deviation = 1.64
Median = 3

b. Mean = 3
Variance = 6.2727
Standard Deviation = 2.5045
Mean Deviation = 1.9048
Median = 3

5. a. weighted average = 2.044

8. a. 12.5
b. 159.17

c. 124.95

d. 10.8

e. 750

9. 3.33, 0.68

6

APL Practice

2. 0.866
0.500
0.577
1.732

3. (b) ∇ CHECK [1⎕]
[1] S←(⌊.5++/÷(0,⍳50)!50) = 2∇
CHECK
YES

4.

Problem	Answer
$F < M$	(~F)∧M
$F \geq M$	F∨~M
$F > M$	F∧~M

5. 19.125

6. (a) 84

(c) 2

7. 166.67

Statistical Applications

1. a. N and M

e. 1 MNTS X
0
2 MNTS X
2.66
3 MNTS X
0
4 MNTS X
15.46

2. a. The sixth element of the series 10 bi .3 is 0.1029.

d. 1−+/5ρ10*BI* .3 or 0.16

4. for P = .5 and K = 4 the probability $f(N)$ = 0.0625

6. 2 POISSON 1
0.1839

12. Mean = 5
Variance = 2.5
Third moment = 0
Fourth moment = 17.5
Like the normal this distribution is bell shaped but less peaked than the normal.

14. Mean = 6
Variance = 2.4
Third moment = ¯0.48
Fourth moment = 16.224
The distribution is skewed to the left (since $V_3 < 0$).

15. a. 6 = 1/1008C

c. 0.055

h. 0.94

17. a. 0.6

d. 1.39

e. 6.11

7

APL Practice

1. A (a) 1 2 3 4 5 6 7 8 9 10

(f) *A B C DA B C D*8

B (d) *A B C DE F*

(e) *A B C DE FE F*

2. (d) MEAN
These data are for part D of Exercise No. 2
⎕:
(7⊛8,3) × ⍳100
54, 921

5.)*WSID*
CLEAR *WS*

6. (b) (+/*DATA*) ÷ ρ*DATA*

8. (d) 83325

11. [3.5] → *EXIT* × ⍳ 1=∧/'*FINISHED*'= 8ρ*DISP*

Statistical Applications

2. a. ¯3.8696E¯17
   b. ¯0.84543

3. a. 0.06 0.22 0.31 0.26 0.1 0.03 0.02
   b. .59 .41

4. *TSX* ← *NORM N*
   [1] *X* ← ? (*N*,30)ρ10
   [2] *SX* ← (((+/X)÷30)−5.5)÷ (8.25÷30)∗.5 ∇

8

APL Practice

1. (a) 7
   12
   (d) 1 *YOU*
   (e) *WTACWTACWT*
   WWWWWWWWWWTAC
2. They must be scalars
3. line [1] (+/*X*) not (+/×)
   line [4] ×+/(*Y*−*MY*) not ××/(*Y*−*MY*)
4. [3]→(*N* ≥ *PR*)/2
7. *W*∘.⌈*K*
11. (c) [12.2][0]*X TTEST Y*; *N*1; *SS*; *X*1; *X*2; *NT*; *C*; *G*; *A*; *N*2; *NO*

Statistical Applications

2. The result of *A TTEST B* is ¯1.15 and the *t*-value from the *t*-table for 90% confidence is ¯1.812. Thus the program was not effective.
3. 0.15
4. a. Computed poisson frequencies are
   223.13 334.7 251.02 125.51
   47.068 14.12 3.53 0.75643
   0.14183 0.023683
5. a. $32.50

9

APL Practice

1. (c) U←*Y*−*A*−*B*×*X*
   (e) *SU*←*S*÷(ρ*X*)−2
   (k) *VYXO*←*XU*×((*XO*−*XB*)∗2)÷ *XSQ*)+÷*N*
5. *F*(20) = !19
   F(3.5) = !2.5

Statistical Applications

1. a. $R^2 = 0.043$
   d. $R^2 = 0.044$
   f. $R^2 = 0.036$
3. Since $F = .507$ for the column means we are unable to establish a dependency on dusting methods.
4. a. F for column means = 10.4, $F(3,12)$ = 3.49 so we reject hypothesis of no difference.
7. Row means $F = 2.3401$
   Column means $F = 0.045685$
   Row & Col. means $F = 1.4223$
   a) *Ho* not rejected
   b) *Ho* not rejected
   c) *Ho* not rejected.

10

APL Practice

2. (a) 2 ¯5
   ¯1 3
   (b) syntax error
   (c) 2 ¯1
   ¯5 3
   (k) 123 78
   44 28

4. Left inverse is given by ⌹ Z
((⌹Z+.×Z)=(⍉Z)+.×⍉⌹ Z
results in
1 1
1 1

6. 2.3 1
¯3.7 1
4.7 1
The rounding of the constants changed the solutions dramatically.

8. a to g result in a matrix of ones proving that the statements are true.

10. $A=A+.\times A$ and $A=(⍉A)+.\times A$ result in matrices of ones.

12. 36.5 and 55.7

Statistical Applications

1. a. $\bar{a}_1 = \frac{15}{5}$, $\bar{a}_2 = \frac{9}{3}$, $var\ a_1 = \frac{1}{5}$, $var\ a_2 = \frac{2}{9}$

2. $\bar{x} = 909.91$, $\bar{y} = 561.12$,
$\hat{a} = {}^{-}39.532$, $\hat{b} = 0.66012$
$t = 69.824$, $R^2 = 0.99449$

3. a. $\bar{x} = 909.91$, $\bar{y} = 46.934$,
$\hat{a} = {}^{-}38.313$, $\hat{b} = 0.093688$
$t = 32.659$, $R^2 = 0.97531$
b. $\bar{x} = 8.9271E5$, $\bar{y} = 46.934$,
$\hat{a} = 2.4623$, $\hat{b} = 4.9817E^{-}5$
$t = 45.556$, $R^2 = 0.98716$

11

APL Practice

4. (b) 1 1 1 0 1\[2]X
(c) 1 0 0 1 ⍀ X
(d) 1, ¯1 44 ↑X

5. C←((0,⍳9)∘.!0,⍳9)×10 10 ρ1 ¯1
then
(C+.×C)×(⍳10)∘.=(⍳10)
gives a matrix of ones.

8. A←2 2ρ1 0 0 2
B←2 2ρ3 0 0 1
(A×B)=B×A
1 1
1 1

10. 1. +/[2]+/[3]ALL
2. ⌈ /[2]+/[3]ALL
3. +/+/+/ALL

Statistical Applications

1. Macro [;4] Regress Macro [;10]
$\bar{x} = 5.3714$, $\bar{y} = 135.76$, $\hat{a} = 44.673$,
$\hat{b} = 16.957$,
$t = 9.837$, $R^2 = 0.78185$.

2. Macro [;4] Regress Macro [;10]
after changing lines [4] and [5] to
[4] SSE←225×(n−2)
[5] V←225
$\bar{x} = 5.3714$, $\bar{y} = 135.76$,
$\hat{a} = 44.673$, $\hat{b} = 1.8179$,
$t = 0.5126$, $R^2 = 0.86654$

12

APL Practice

1. The required number is the mean of L

5. Q is a list of N random numbers from 1 to K

6. (a) (⍳100)∘.*⍳10
(c) (⍳100)∘.*.1×⍳10
(f) S←30 30ρ(.2*0,⍳29),.2 for ρ=.2

10. (a) 11.038
(b) 11.338, 11.15, 11.038, 11.225

13. (b) a row of ones
(c) you obtain c
(g) a matrix whose rows is either ones or not c.

13

Statistical Applications

1. a. Beta coefficients are 2.6706 and $^-129.42$
   their t ratios are 32.209 and $^-1.7648$

3. Coefficients and corresponding t-ratios

a^*	a_1	a_2
1.0223	0.42378	0.49806
(0.027089)	(18.425)	(2.3033)

4. Coefficients and corresponding t-ratios

a_0	a_1	a_2
2730.8	−1.5229	50.688
(2.9563)	(−2.1977)	(6.3059)

5. b. $\gamma = .9$, $b_1 = {}^-.14$, $b_2 = .37$,
   $b_0 = 57.67$

7. t ratio for $C_3 = {}^-1.9948$
8. t-statistics
   −2.3658, 51.987, 3.9462,
   −4.5486, 3.3161, 0.34169
   F = 25.888

9. a. F statistic = 5.1094 with 30 and 120 degrees of freedom.
   b. F statistic = 2.3604 with 30 and 120 degrees of freedom.

10. $a_0 = 7.7956$ and $a_1 = {}^-0.32873$
    Elasticity = −.32873

14

Statistical Applications

3. t-statistic for $a_2 = 2.35$ thus the H_0 is rejected.
   F statistic = 197.7 and H_0 is also rejected

5. a. equilibrium .

6. b. *TTest* 1
   $^-0.23881$, 0.79786, 0.79786

7. Beta coefficients = $^-569.26$, 2.6626, 19.659
   t-ratios = $^-6.4608$, 204.67, 7.0156

9. c. $a_0 = 2.7769$, $a_1 = 0.80121$

11. $a_0 = {}^-10.298$, $a_1 = 1.1005$

12.

coefficients	t values
$1\text{-}\gamma = 0.55511$	5.9576
$\gamma_b = 0.30999$	5.2282
$\gamma_a = {}^-22.376$	$^-3.316$

Using $Y_{1978} = 1{,}385.1$
and $C_{1977} = 857.7$
we obtain
$C_{1978} = {}^-22.376 + .31(1{,}385.1) + .56(857.7) = 883.11$
which is close to the actual value of 891.2.

Index

Symbol Index